THE SCIENTIST IN THE CRIB

ALSO BY ALISON GOPNIK AND ANDREW N. MELTZOFF

Words, Thoughts, and Theories

Praise for
The Scientist in the Crib

"Fascinating. . . . It isn't until you read *The Scientist in the Crib* alongside more conventional child-development books that you begin to appreciate the full implications of its argument."

—MALCOLM GLADWELL,
THE NEW YORKER

"*The Scientist in the Crib* is a triumph, a clearheaded account of the kinds of things that go on in the heads of young children. . . . [This book] speaks in the voice of intelligent parents talking to other intelligent parents—witty, rather personal, and very well informed."

—*NEW YORK REVIEW OF BOOKS*

"Meticulously researched, combining charm and erudition, humor and humanity, *The Scientist in the Crib* . . . should be placed in the hands of teachers, social workers, therapists, policymakers, expectant parents, and everyone else who cares about children. . . . This book is full of enchanting revelations."

—*WASHINGTON POST*

"[An] excellent book. . . . It should be of interest to anyone curious about the human mind and its origins."

—*CHICAGO TRIBUNE*

"[*The Scientist in the Crib*] covers a remarkable period of research that has revolutionized the field of cognitive development. . . . [The authors] show how an understanding of adult/baby interactions can open a new window on human evolution."

—*SAN FRANCISCO CHRONICLE*

"A brilliant . . . book by three superstars in the study of children's developing minds."

—GARRY WILLS, PH.D., adjunct professor of history, Northwestern University, and Pulitzer Prize winner

"This is a terrific book—a page-turner, in fact. I couldn't wait to see what was going to happen next. It is a treasure trove of cutting-edge science combined with beautiful writing, warmth, humor, and illuminating anecdotes that only these scientist-parents could provide. It is a must for all parents and anyone working with infants and young children."

—MICHAEL B. ROTHENBERG, M.D., coauthor of *Dr. Spock's Baby and Child Care,* fifth and sixth editions

"This book is at once a masterful synthesis of the latest findings about the minds of children and a provocative argument that young children resemble practicing scientists. Few books about human development speak so eloquently to both scholars and parents."

—HOWARD GARDNER, PH.D., author of *Intelligence Reframed: Multiple Intelligences for the 21st Century*

"Finally, a book on child development that is both authoritative and really fun to read! Professionals, students, and the public at large can learn from it. In short, a brilliant book."

—JOHN H. FLAVELL, PH.D., professor of psychology, Stanford University

"This book provides a wealth of insight into our youngest children. Written by leading scientists in early child development, it shows us how children discover the world. No voyage of discovery was ever more important than this one. To read this book is a joy and an illumination of fundamental questions about human nature."

—DAVID A. HAMBURG, M.D., president emeritus, Carnegie Corporation of New York

THE SCIENTIST IN THE CRIB

What Early Learning Tells Us About the Mind

Alison Gopnik, Ph.D.
Andrew N. Meltzoff, Ph.D.
Patricia K. Kuhl, Ph.D.

Perennial
An Imprint of HarperCollins*Publishers*

A hardcover edition of this book was published in 1999 by William Morrow and Company, Inc.

HarperCollins books may be purchased for educational, business, or sales promotional use. For information please write: Special Markets Department, HarperCollins Publishers Inc., 10 East 53rd Street, New York, NY 10022.

First Perennial edition published 2001.

Designed by Oksana Kushnir

The Library of Congress has catalogued the hardcover edition as follows:

Gopnik, Alison.
The scientist in the crib : minds, brains, and how children learn / Alison Gopnik, Andrew N. Meltzoff, Patricia K. Kuhl.
p. cm.
Includes index.
ISBN 0-688-15988-5
1. Cognition. 2. Cognition in children. 3. Learning, Psychology of. I. Meltzoff, Andrew N. II. Kuhl, Patricia K. (Patricia Katherine). III. Title.
BF311.G627 1999
155.4'13—dc21 99-24247
CIP

ISBN 0-688-17788-3 (pbk.)

02 03 04 05 WB/RRD 10 9 8 7 6 5 4

For all our children

PREFACE AND ACKNOWLEDGMENTS

Scientists and cribs? We wrote this book to show that scientists and cribs, and the children in them, belong together. For the last thirty years scientists like us have been looking in cribs—and in playpens and nurseries and preschools. There have been hundreds of rigorous scientific studies that tell us how babies and young children think and learn. These studies have revolutionized our ideas about babies and young children, and about the nature of the human mind and brain. They have also helped answer profound and ancient philosophical questions. We can learn as much by looking in the crib and the nursery as by looking in the petri dish or the telescope. In some ways we learn more—we learn what it means to be human.

In this book we tell the story of the new science of children's minds. This story should be important to everyone who is interested in the mind and the brain. It's a central part of the new discipline called cognitive science. Cognitive science has united psychology, philosophy, linguistics, computer science, and neuroscience. New scientific insights often come from unexpected and even humble places, and some of the most important insights in cognitive science have come from the crib and the nursery. Understanding children has led us to understand ourselves in a new way.

Scientists and children belong together in another way. The new research shows that babies and young children know and learn more about the world than we could ever have imagined. They think, draw conclusions, make predictions, look for explanations, and even do experiments. Scientists and children belong together because they are the best learners in the universe. And that means that ordinary adults also have more powerful learning abilities than we might have thought. Grown-ups, after all, are all ex-children and potential scientists.

We hope this book will demonstrate that scientists and children belong together in still other ways. Parents are deeply, even passionately interested in children, or at least in their children. But parents find that their interest in children is treated differently from their interest in science. Books about science assume that their readers are serious, knowledgeable, intelligent, sophisticated adults who simply want to know about the things they care about. But books about babies and children are almost all books of advice—how-to books. It's as if the only place you could read about evolution was in dog-breeding manuals, not in Stephen Jay Gould; as if, lacking Stephen Hawking's insights, the layman's knowledge of the cosmos was reduced to "How to find the constellations." How-to books can be enormously useful, but they shouldn't be the only place parents can learn about something they care about as much as they care about children.

We hope this book will help fill that gap. The science of babies' minds should hold a special fascination for people who live with babies and young children every day. The picture of children that emerges is at once surprisingly familiar and surprisingly unfamiliar. Parents who read this book should find themselves feeling both the shock of recognition and the shock of the new.

There is yet another reason that scientists and children belong together. Everyone should be interested in understanding children because the future of the world, quite literally, depends on them. Recently there has been more and more recognition of that fact. But getting public policies about children right depends on getting the science right. The political sound bites and op-ed-page pieces are inevitably simplified. If citizens and voters are going to make the right political decisions about children, they need to understand what science tells us (and what it doesn't).

In writing this book we've faced the usual problems of scientists trying to explain their research. Science is elegant and orderly. But it is also messy, noisy, complicated, and invariably embroiled in controversies and debates. We've tried to present what we think are the most interesting experiments, conclusions, ideas, and speculations, but we couldn't possibly reflect the entire field in all its diversity and complexity. We've tried to indicate when we are talking about our own views and when we are talking about ideas that are generally accepted in the field, and to indicate the many questions that remain unanswered.

The new science of development, like any science, depends on the cumulative efforts of literally thousands of scientists. It would be impossible to acknowledge them all in the text, and anyway, it would make readers feel as if they were at a party where everyone kept talking about people they didn't know. We have tried at least partly to remedy this by including detailed and extensive source notes and a bibliography at the end of the book. They are designed to give scientific references for our factual claims and to point to the best and clearest accounts of the central ideas.

Part of the message of this book is that children can do so much because they have the help of people who care about them. This is even more true of authors. This book depends on an entire generation of scientists who showed that babies had minds and that studying those minds was important and valuable. It also depends on the thousands of parents and children who generously and enthusiastically participated in the research.

Our own ideas and research were supported by grants from the National Science Foundation and the National Institutes of Health (NSF9213959, HD22514, HD18286, HD34565, and DC00520). We have been generously supported by the Department of Psychology, the Institute of Human Development, and the Institute of Cognitive Studies at the University of California at Berkeley, and also by the Department of Psychology, the Department of Speech and Hearing Sciences, and the Center on Human Development and Disability at the University of Washington. We are also grateful to our colleagues and students at both universities.

John Campbell and Danny Povinelli read and commented on drafts of this book, and we are very grateful. We had unusual genetic luck: Adam Gopnik, a generous brother as well as a masterly writer, made especially helpful comments and suggestions, and Julian Meltzoff provided the wisdom of a father as well as a scientist and sympathetic reader. Katinka Matson, our agent, helped make this project a reality. Toni Sciarra, at Morrow, has been unfailingly enthusiastic and helpful, and an exemplary editor. Thanks also go to Keith Moore for years of collaborative work, and to Craig Harris, Calle

Fisher, and Erica Stevens for assisting with the research and the process of preparing the final manuscript.

Authors always end by acknowledging their families. But for this book that acknowledgment takes on a special importance. Contemplating childhood is especially satisfying for us because our own childhood experiences were so luminous and so full of parents, brothers, and sisters who cared about us and taught us at the same time. We are deeply grateful to Irwin and Myrna Gopnik, and to Adam, Morgan, Hilary, Blake, and Melissa; to Julian and Judith Meltzoff and Nancy; and to Joe and Susan Kuhl and Delphine, Donna, Benno, and Shirley.

Combining science and children hasn't been just the project of this book, it has been the most important and most profoundly satisfying project of our lives. Andy and Pat are married and deeply grateful for each other, but Alison has to express the deepest gratitude to her husband, George Lewinski, for his help in that project. This book simply could not have been written without our children, Katherine Meltzoff and Alexei, Nicholas, and Andres Gopnik-Lewinski. It is dedicated to them, and to all the others.

CONTENTS

Thou, whose exterior semblance doth belie
Thy Soul's immensity;
Thou best Philosopher, who yet dost keep
Thy heritage, thou Eye among the blind,
That, deaf and silent, read'st the eternal deep,
Haunted for ever by the eternal mind,—
Mighty prophet! Seer blest!
On whom those truths do rest,
Which we are toiling all our lives to find . . .
Thou little Child, yet glorious in the might
Of heaven-born freedom on thy being's height . . .

—*Wordsworth*
"Ode: Intimations of Immortality from
Recollections of Early Childhood"

THE SCIENTIST IN THE CRIB

CHAPTER ONE

Ancient Questions and a Young Science

Walk upstairs, open the door gently, and look in the crib. What do you see? Most of us see a picture of innocence and helplessness, a clean slate. But, in fact, what we see in the crib is the greatest mind that has ever existed, the most powerful learning machine in the universe. The tiny fingers and mouth are exploration devices that probe the alien world around them with more precision than any Mars rover. The crumpled ears take a buzz of incomprehensible noise and flawlessly turn it into meaningful language. The wide eyes that sometimes seem to peer into your very soul actually do just that, deciphering your deepest feelings. The downy head surrounds a brain that is forming millions of new connections every day. That, at least, is what thirty years of scientific research have told us.

This book is about that research. What are these deeply familiar yet surprisingly strange creatures we call children really like? Of course, human beings have always wondered, pondered, and even agonized about their children. But most of the time, the questions people ask are practical. Some are immediate, questions about how to get them to eat more or cry less. Some are long-term, questions about how to turn them into the right kind of grown-ups. These are important questions, crucial for the survival of any civilization (not to mention any parent), but we won't have very much to say about them. This book won't tell you how to make babies easier or smarter or nicer, or how to get them to go to sleep or to Harvard. There are lots of books that do that, or anyway say they do, right between the cooking and house-repairs sections in your local bookstore. Our questions are both harder and easier than the practical questions. We want to understand children, not renovate them.

While the purported answers to the practical questions fill volumes, all of us who have lived with babies and young children, or even just looked at them, have found ourselves asking deeper questions. We decided to become developmental psychologists and study children because there aren't any Martians. These brilliant beings with the little bodies and big heads are the closest we can get to a truly alien intelligence (even if we may occasionally suspect that they are bent on making us their slaves). Babies are fascinating, mysterious, and just plain weird. Watch awhile. A three-month-old catches sight of the stripes on a shopping bag and follows it carefully as her father carries it around the room, staring with intense cross-eyed concentration. A one-year-old visiting the zoo points at the elephant and says triumphantly and with great certainty, "Doggie!" A "terrible two-year-old" turns toward the expressly forbidden switch of the computer and slowly, deliberately, watching his mother every moment, erases the day's work. As we change diapers and wipe noses, all of us, no matter how preoccupied, find ourselves exclaiming, "What's going on in that little head of hers? Where on earth did he get *that* from?"

Developmental psychologists have had the luxury of asking those questions systematically and even getting answers to them. We're actually starting to understand what's going on in that little head of hers and where on earth he got that from.

Studying babies is full of fascination in its own right. But developmental research also helps answer a more general, deep, and ancient question, not just about babies but about us. We human beings, no more than a few pounds of protein and water, have come to understand the origins of the universe, the nature of life, and even a few things about ourselves. No other animal, and not even the most sophisticated computer, knows as much. And yet every one of us started out as the helpless creature in the crib. Only a few tiny flickers of information from the outside world reach that creature—a few photons hitting its retinas, some sound waves vibrating at its eardrums—and yet we end up knowing how the world works. How do we do it? How did we get here from there?

The new research about babies holds answers to those questions, too. It turns out that the capacities that allow us to learn about the world and ourselves have their origins in infancy. We are born with the ability to discover the secrets of the universe and of our own minds, and with the drive to explore and experiment until we do. Science isn't just the specialized province of a chilly elite; instead, it's continuous with the kind of learning every one of us does when we're very small.

Trying to understand human nature is part of human nature. Developmental scientists are themselves engaged in the same enterprise and use the same cognitive tools as the babies they study. The scientist peering into the crib, looking for answers to some of the deepest questions about how minds and the world and language work, sees the scientist peering out of the crib, who, it turns out, is doing much the same thing. No wonder they both smile.

The Ancient Questions

How can we know so much when our senses are so limited? This problem—the problem of knowledge—is one of the oldest and most profound problems of philosophy. The branch of philosophy called epistemology is devoted to it. Three versions of the problem are especially important and puzzling to grown-ups and children alike. We'll call them the Other Minds problem, the External World problem, and the Language problem. The new developmental psychology helps answer all three.

Take a perfectly ordinary event. Every Sunday night, we sit around the dinner table. We serve up healthy leek and potato soup (which must be eaten before you get dessert), pass the salt and pepper, butter the bread, push our chairs back from the big wooden table. We laugh, fight, and tease one another. One of the big brothers invariably makes a rude joke at the expense of the little brother, who is hurt and demands an apology. No experience could be more banal, more domestic, more comfortable and familiar. Except that, actually, we don't experience any of this at all.

All that really reaches us from the outside world is a play of colors and shapes, light and sound. Take the people around the table. We seem to see husbands and wives and friends and little brothers. But what we really see are bags of skin stuffed into pieces of cloth and draped over chairs. There are small restless black spots that move at the top of the bags of skin, and a hole underneath that irregularly makes noises. The bags move in unpredictable ways, and sometimes one of them will touch us. The holes change shape, and occasionally salty liquid pours from the two spots.

This is, of course, a madman's view of other people, a nightmare. The problem of Other Minds is how we somehow get from this mad view to our ordinary experience of people. Why is it we don't see skin-bags but husbands and wives and children—people with thought and feelings, beliefs and desires like ours, including wounded pride that demands apologies?

We don't even really see the things in the room, either. The brown, bounded shape we think of as the table perpetually changes its form as we move around it. The apparently solid three-dimensional spoons and pepper mills are really just flat surface images on our eyes. The feel of the spoon in our hands is quite different from the shape we see. The surface of the table is full of discontinuities: white holes where it is hidden by plates and bowls. The soup changes its form even more radically as it moves from tureen to spoon to mouth until we lose sight of it altogether and only feel the warmth in our throats. We seem to know about a world of objects with properties that are quite independent of us, a world of tables and spoons and healthy soup. But all we experience directly is an endlessly changing chaotic flow of sensations. This is the External World problem.

The problem is perhaps worst of all when we turn to the sounds that come out of the holes in the skin-bags. Sit in a café in a foreign city. Suddenly you'll realize that the thoughts and jokes and apologies that float so unself-consciously around the dinner table are really a blazingly fast succession of finely modulated noises, each just barely different from the last. Each word is actually nothing but a transitory whisper of a disturbance in the air that lasts for milliseconds until it's replaced by the next. The most sophisticated computers can barely decode continuous speech spoken by a single person in a calm voice. Yet for us the words are completely transparent: we experience only the ideas of the people who speak them. We can hear a sentence spoken by a little boy with a soul full of excited indignation and a mouth full of soup and turn it effortlessly into a thought. This is the problem of Language.

The sensitive three-year-old little brother at the table can do all these things, too. He experiences his brothers teasing him, not skin-bags moving. He sees tables and spoons and healthy soup, not undifferentiated colors and shapes. And he immediately understands the significance of the rude joke and the apology that are actually no more than the most fleeting vibrations. How can he do it?

Baby 0.0

The modern answer to this question is that babies are a kind of very special computer. They are computers made of neurons, instead of silicon chips, and programmed by evolution, instead of by guys with pocket protectors. They take input from the world, the flickering chaos of sensations, and they (and therefore we) somehow turn it into jokes, apologies, tables, and spoons. Our job as developmental psychologists is to discover what program babies run and, someday, how that program is coded in their brains and how it evolved. If we could do that, we would have solved the ancient philosophical problems of knowledge in a scientific way.

Thinking of babies' minds as computers made of neurons and programmed by evolution makes us see babies differently, but it also makes us see computers and neurons and evolution differently. The baby computers must be much more powerful than even the most impressive product of Silicon Valley. Bill Gates's little daughter has already solved problems that Bill, with all his billions, is still unsuccessfully trying to crack. The new developmental research tells us that Baby 0.0 must have some pretty special features.

First, it must already have a great deal of knowledge about the world built into its original program. The experiments we will describe show that even newborns already know a great deal about people and objects and language. But more significant, babies and children have powerful learning mechanisms that allow them to spontaneously revise, reshape, and restructure their knowledge. This is, notoriously, the great weakness of existing computers. They are terrific at solving well-defined problems, they are not so hot at learning, and they are really awful at spontaneously changing how they learn. Finally, the babies have the universe's best system of tech support: mothers. Grown-ups are themselves designed to behave in ways that will allow babies to learn. This support plays such a powerful role in the babies' development, in fact, that it may make sense to think of it as part of the system itself. The human baby's computational system is really a network, held together by language and love, instead of by optic fiber.

Studying babies also makes us think about the brain in a new way. People often seem to split the human mind into two parts: a "natural" neurologically determined part that is shaped by evolution and a "cultural" socially determined part that is shaped by learning. Studying babies makes us realize how deeply misguided these oppositions are. They aren't just misguided in the obvious sense that there is an interaction between nature and nurture or that there is a little of both. They are misguided in a much deeper sense. *Everything* about our minds is the result of what happens in our brains, from the most automatic mechanisms that govern our breathing to the most refined, culturally elaborated details of wedding etiquette and existential angst. That means, though, that the brain must be profoundly flexible, sensitive, and plastic, and be deeply influenced by events in the outside world. A handful of genes couldn't predetermine the billions of specific neural connections that make up an adult brain. And, as we'll see, the more we learn about the brain, the more flexible, sensitive, and plastic it seems. This is partly because we have only very recently started analyzing live brains instead of dissecting dead ones; living things generally look more active than dead things.

Just as everything about our minds is caused by our brains, everything about our brains is ultimately caused by our evolutionary history. That means, though, that evolution can select learning strategies and cultural abilities just as it selects reflexes and instincts. For human beings, nurture *is* our nature. The capacity for culture is part of our biology, and the drive to learn is our most important and central instinct. The new developmental research suggests that our unique evolutionary trick, our central adaptation, our greatest weapon in the struggle for survival, is precisely our dazzling ability to learn when we are babies and to teach when we are grown-ups.

In fact, we've proposed a more specific version of this general evolutionary picture. If you look at a wide range of animal species, a few evolutionary characteristics seem to go together. Animals with a relatively large cortex, behavioral flexibility, and cognitive complexity (what we anthropomorphically think of as smart animals, although the cockroaches might give us an argument) also tend to share other traits. These include eating a wide variety of foods, having sex lots of different ways, being polygamous, living in lots of different places, and—most important for our purposes—having a long period of immaturity. Passing quickly over polygamy, we human beings have all these traits in spades (particularly in Berkeley).

That long period of immaturity, childhood, is a puzzle. Why leave the young so helpless for so long, and why require the adults to invest so much time and energy in protecting them? One idea raised by the eminent psychologist and educator Jerome Bruner is that that period of protected immaturity allows children to learn about their specific physical environment (we humans can survive in more different environments, including outer space, than any other creature). Even more significant, it allows children to learn about their specific social environment (we organize ourselves into more different kinds of social groups than any other creature). Other species survive by having elaborately developed instincts that are exquisitely adapted to their particular ecological niche. We survive by being able to learn how to behave in almost any ecological niche, and by being able to construct our own niches.

If this is our evolutionary strategy, it makes sense to have babies who are brilliantly intelligent learners and grown-ups who are deeply devoted to helping them learn. That may be why we also have babies who are utterly helpless and grown-ups who are devoted to keeping them alive. The advantage of learning is that it allows you to find out about your particular environment. The disadvantage is that until you do find out, you don't know what to do; you're helpless. We may have two evolutionary gifts: great abilities to learn about the world around us and a long protected period in which to deploy those abilities.

We've even argued that our otherwise mysterious adult ability to do science may be a kind of holdover from our infant learning abilities. Adult scientists take advantage of the natural human capacities that let children learn so much so quickly. It's not that children are little scientists but that scientists are big children.

Of course, evolutionary arguments about the mind are necessarily speculative. What we do know for sure, or as sure as anything can be in science, is that babies *are* brilliantly intelligent learners and that grown-ups *are* devoted to helping them learn. That's what we'll show in the next few chapters.

The Other Socratic Method

The problems of knowledge—the problems of Other Minds, the External World, and Language—are, literally, ancient questions. The new idea is that studying very young children and babies can help answer them. Since children have always been around, why did it take so long before scientists looked at them?

Curiously enough, the idea that children were important was there at the very beginning. One of the first and most famous formulations of the problem of knowledge, and the first attempt at an answer, comes in the Socratic dialogue called the *Meno.* Socrates and his friends, drinking wine at Meno's house, are considering how we can understand something as abstract as virtue when we have no direct experience of it. Socrates' answer is that we did not learn about virtue, or any other abstract concept, from our experience. We must have known about it in the first place. Socrates thought we remembered it from a past life. Contemporary versions of Socrates' argument would say that it is in our genetic code.

Every first-year philosophy student learns about Socrates' argument. What that first-year student doesn't learn, though, is that it isn't just an abstract logical argument. It's actually based on a kind of experiment, a piece of empirical scientific investigation. The most important person in the *Meno* isn't Meno or Socrates or any of the aristocrats but an anonymous child, the slave boy who pours the wine. The *Meno* is both the first discussion of the problem of knowledge and the first recorded developmental psychology experiment.

Socrates moves from abstract concepts like virtue to the even more abstract concepts of geometry. He takes the slave boy, an uneducated child, through the steps of a geometric proof. The boy acknowledges the truth of each step and ends up proving the theorem. Socrates concludes that because the boy, who has had no experience of geometry, can do this, he must already know the proofs of geometry without being aware of them.

This is a pretty stunning conclusion now, but it was even more stunning then. Euclid worked at the Academy in Athens, and his proofs were being formulated around the same time that the *Meno* was written. Geometry was the most exciting cutting-edge science of Socrates' day. It's as if we said that children really know Andrew Wiles's proof of Fermat's theorem.

The new research shows that Socrates' stunningly counterintuitive idea was right: even tiny babies and uneducated children must know much more than we think. That's the first element in the modern answer to the problem of knowledge. But, as we'll see, it isn't the whole answer.

The Great Chain of Knowing

For the next 2,500 years philosophers talked and wrote at (sometimes agonizing) length about the problem of knowledge, but no one tried Socrates' method again. No one tried to solve the problem by talking to children and finding out what they knew. (An English philosopher once explained to us that while he had, of course, seen children about, he had never actually spoken to one.)

In fact, "children's knowledge" seemed like a contradiction

in terms. The dominant view was that children were essentially defective adults. They were defined by the things they didn't know and couldn't do. Of course, there was something right about this; we can all see that there are many things children don't know and can't do, and in our role as parents we focus on remedying those defects. But this picture of children was also deeply misleading.

It was a picture of a sort of "great chain of knowing," with babies at one end and philosophers at the other. According to this picture, children (and "primitive" peoples and women) had characteristics that were the very opposite of the characteristics of reason, science, and civilization. They were intuitive and not rational, natural and not civilized, driven by passions rather than guided by plans. Their minds were captured by appearances, superstitious and magical. The younger children were, the further removed they were from knowledge. Newborn babies were, literally, nothing at all. In the famous metaphor of the seventeenth-century philosopher John Locke, they were "tabulae rasae," blank tablets.

There was also an opposing view, articulated most clearly by the Romantic poets and philosophers of the early nineteenth century. One of its most eloquent expressions was Wordsworth's great poem "Ode: Intimations of Immortality from Recollections of Early Childhood," part of which prefaces this book. The Romantic view was that children (and "primitive" peoples and women) had a special kind of knowledge just because they were so ignorant. Children's knowledge was like poetry but not like science. The Romantics thought that children's experience had a kind of clarity and intensity precisely because it was uncorrupted by adult conceptions. And, of course, there was something right about that, too. No one could look at children, or trust their own recollections of childhood experience, without glimpsing extraordinary powers of perception and imagination.

But even the Romantic view of children shared the central assumptions of the dominant view. The Romantics actually agreed that children were intuitive, irrational, uncivilized, governed by passion, and at the furthest remove from science. They just thought that those were all good things rather than bad, powers and not defects, sources of knowledge and not impediments to it.

Those two views live on, and so do the assumptions that underlie them. There is still supposed to be a deep split between scientific, cultivated, rational ways of knowing the world and intuitive, natural, emotional ways of knowing. And children (and "primitive" peoples and women) are still assumed to be the exemplars of intuition rather than science, and passion rather than reason. The debate is still about which side you think you ought to root for.

The new developmental research shows that this historical consensus about children was just plain wrong. Children are not blank tablets or unbridled appetites or even intuitive seers. Babies and young children think, observe, and reason. They consider evidence, draw conclusions, do experiments, solve problems, and search for the truth. Of course, they don't do this in the self-conscious way that scientists do. And the problems they try to solve are everyday problems about what people and objects and words are like, rather than arcane problems about stars and atoms. But even the youngest babies know a great deal about the world and actively work to find out more.

That undermines the entire picture of the great chain of knowing. Women and people from other cultures have, after all, at least escaped the negative implications of being "childlike." (Nowadays it's okay to think women and people from other cultures are intuitive and natural only if you take the positive, Romantic view.) But if even children themselves aren't "childlike," the whole picture collapses. There are no savages, noble or otherwise, and there are no "children of nature," not even among children. There are only human beings, children and grown-ups, women and men, hunter-gatherers and scientists, trying to figure out what's going on.

The curious thing is that the consensus about children never was supported by any systematic evidence. Science is supposed to go beyond our everyday assumptions, to test what we all think we know. But no one set out to see what children really did or didn't know about the world, what they began with and what they learned. Bertrand Russell made a nice remark about Aristotle's claim that women had fewer teeth than men. The surprising thing wasn't so much that Aristotle was wrong but that all he had to do to find out he was wrong was ask Mrs. Aristotle to open her mouth, and count. Children, after all, are all around us; we don't need expeditions to distant continents or high-tech laboratories to observe them. All we had to do was ask them to open their mouths, and listen. For 2,500 years, nobody did.

Piaget and Vygotsky

The modern revival of the other Socratic method began only in the 1920s. It happened in two obscure corners of the world: quiet, dull, prosperous, peaceful Geneva and war-wracked, famine-stricken, tyrant-ridden Moscow. Jean Piaget was a precociously brilliant biologist and naturalist. He published his first academic paper, "On an albino sparrow," when he was only ten years old. By the age of twenty he had published dozens of papers on biology, mostly about mollusks. But he had also read Kant's *Critique of Pure Reason.* Before he was thirty, he was appointed head of the J. J. Rousseau Institute at the University of Geneva, where he remained for the rest of his very long life.

Piaget wanted to explain the classical philosophical problems of knowledge. But unlike earlier philosophers he wanted an explanation that would be linked to biology. At the turn of the century, advances in biology were beginning to make us realize that the mind must somehow be instantiated in the brain. Human knowledge itself must be a natural, biological phenomenon. Piaget wanted to find a link between Kant and the mollusks, between epistemology and biology. His great insight was that studying the development of human children was one way to do this.

In the thirties Piaget began to record the lives of his own three infant children, Jacqueline, Lucienne, and Laurent. There have been baby diaries before and since, but there is nothing like the Piaget diaries. They record in minute, crystalline detail the significant patterns in the apparently formless behavior of very young babies. Moreover, they do so day by day and even moment by moment, so that each observation becomes part of a larger unfolding history. By reading Piaget's books we know Jacqueline's, Lucienne's, and Laurent's babyhoods more intimately than we know those of our own children. (It was deeply uncanny to meet them, pleasant people our parents' age, at the Piaget centenary in 1996.) The observations of the babies are embedded in an alternately impenetrable and brilliantly insightful theoretical apparatus. When we remember that all this was done without any recording devices other than keen observation, pen, and paper, in the intervals of a demanding academic job, it becomes almost inconceivable that one man could have accomplished it. And, as a matter of fact, we know now that one man didn't accomplish it. Valentine Piaget, Piaget's wife and the children's mother and herself a psychologist, was actually responsible for much of the observation. It's a pretty impressive accomplishment even for two geniuses.

The Piagets, like Socrates, saw that very young children already knew much more about the world than anyone would have thought. But the great and original contribution of their work to the modern view was rather different. In the *Meno*, Socrates is constantly assessing whether the child understands geometry the same way an adult does. Socrates' child always says, "Yes, O Socrates." Piaget also asked children if they saw the world the same way adults did. But Piaget's children said, time and again, *"Non, Monsieur Piaget."* Piaget showed that babies' view of the world was as complex, and as highly structured, as the adult view. And he showed that babies were searching for the truth about the world around them. But he also argued that the babies' view was profoundly, qualitatively different from the adult view. Like adults, babies had systematic ideas about other people, the world, and language, but their ideas were different from ours and often very peculiar. Babies seemed to think, for example, that objects just stopped existing when they were hidden and that there were no boundaries between themselves and others.

Piaget concluded that babies aren't just born in possession of adult knowledge, either from a past life or from DNA. Instead, Piaget thought that children must have powerful learning mechanisms that allow them to construct new pictures of the world, pictures that might be very different from the adult picture. When we learn about the world, when we do science, for example, we don't just hit the right answer once and for all. Rather, there is a very gradual unfolding sequence of corrected errors, expanded ideas, and revised misconceptions as we approach more and more nearly to the truth. That was what the Piagets saw as they watched their babies make their way through infancy.

But Piaget also thought that learning was just as rooted in biology as any innate idea in the genetic code. He often used the metaphor of digestion: babies' minds assimilated information the way babies' bodies assimilated milk. For Piaget, learning was as natural as eating. This idea is the second element in the new developmental science.

The third element in the modern view came from a source quite as unlikely as a Swiss crustaceans expert and his wife. Lev Vygotsky was part of the great intellectual maelstrom of 1920s Russia. He was a literary critic, a doctor, and a consumptive. Like Piaget he wanted to reconcile psychology and biology. But his interest in language and thought was also related to the great political questions of the day. He was a fervent Marxist, and he wanted to know how societies shaped the minds of the people in them.

In 1930s Russia political speculation was even more unhealthy than tuberculosis. The great Russian neurologist Alexander Luria was one of Vygotsky's students. Vygotsky thought literacy might have a profound effect on cognition and perception. He sent Luria to the far east of Russia to test whether the illiterate Tatars experienced certain perceptual illusions. Luria, wildly excited by his results, couldn't wait for the Trans-Siberian Railroad journey back and telegraphed Vygotsky, "Tatars have no illusions." He was immediately arrested; there was only one subject about which Tatars could have no illusions. Luria decided to leave developmental psychology and became a military brain surgeon at the front—it was safer.

Vygotsky himself avoided the purges only by dying young, at thirty-eight. Like Piaget he began his lab when he was still in his twenties, but by then he and his students knew he would soon be dead. For a little less than a decade he was furiously productive, churning out half-reported experiments and unpublished manuscripts, dictated toward the end as he grew too weak to hold a pen. A year after his death Stalin issued an official decree specifically outlawing developmental psychology. (It was a dry run for the more famous decree outlawing evolutionary biology.) The decree was still in force in the 1970s. Most of Vygotsky's students were imprisoned for conducting "bourgeois research."

Vygotsky saw that grown-ups play a crucial role in what children know. For most of us immersed in the practical job of child-rearing, that seems obvious. Our own presence seems self-evidently to be the most determining factor in our children's lives, for good or ill. We are understandably preoccupied with what we do or don't do and how that shapes our children. But the enormous fact of our relation to our children gives us a limited and inaccurate view of them, just as overpowering romantic love, which is the best analogy to parenthood for most of us, blinds us to the true character of our lovers and of love itself. When you read Piaget's diaries, that huge, adoring parental ego is strikingly, even spookily, absent. In fact, that's precisely what gives Piaget's work its great strength. By eliminating himself from the picture, he got a much clearer view of the child's mind than anyone before him. But, of course, adults, and parents especially, are an inescapable fact in children's lives, and by deemphasizing this Piaget missed something important.

Vygotsky saw that adults, and especially parents, were a kind of tool that children used to solve the problem of knowledge, in contrast to our—probably necessary—parental megalomania. Vygotsky noticed, for example, how adults, quite unconsciously, adjusted their behavior to give children just the information they needed to solve the problems that were most important to them. Children used adults to discover the particularities of their culture and society.

But Vygotsky also thought that the adult influence on children's minds was fundamentally biological, a part of our basic human nature. He emphasized the role of language. Language is a natural, biological, and unique feature of human beings, but it is also the medium by which we transmit our cultural inventions. Just as Piaget saw that learning was innate, Vygotsky saw that culture was natural.

In spite of the enormous differences in their theoretical approaches and characters, the tormented, dying, literary Russian and the serene, patriarchal, naturalizing Swiss had two things in common. First, they both developed a method that involved close, detailed observation of babies and young children in natural settings, often over a long period of time. Their conclusions were closely related to those observations. The fact that their theories were rooted in that solid empirical work was, as much as anything, what led them eventually to prevail. Second, their work was almost completely ignored for the next thirty years.

The theories that did dominate psychology, especially in America, were Freudianism and the behaviorism of psychologists like B. F. Skinner. Both theories had lots of things to say about young children. But like Aristotle with the teeth, neither Freud nor Skinner took the step of doing systematic experiments with children or babies. Freud largely relied on inferences from the behavior of neurotic adults, and Skinner on inferences from the behavior of only slightly less neurotic rats. And like the philosophers, Freud and Skinner got the developmental story wrong, too. Freud saw children as the apotheosis of passion, creatures so driven by lusts and hungers that their most basic perceptions of the world were deeply distorted fantasies. Skinner's view was that children were the ultimate blank tablets, passively waiting to be inscribed by reinforcement schedules.

The New View: The Computational Baby

The second and (knock on wood) permanent revival of developmental science had to wait for the late 1960s. Changes in the scientific zeitgeist are always a bit mysterious, but we can point to several quite different factors that suddenly made Piaget and Vygotsky relevant again. Some of it was sociological. For centuries, children were part of women's realm and therefore deemed unworthy of serious scientific interest. So long as men dominated academia, developmental psychology was inevitably marginalized. At Berkeley, for example, a number of renowned women developmental psychologists worked in research institutes; none of them was ever offered a position in the university proper. (In fact, until 1973 there were no women at all in the Berkeley psychology department.) The work that did go on often was in the context of practical issues in education and child-rearing. One male colleague discovered in the sixties at Cornell that he had to take a home economics degree in order to study developmental psychology. The advent of women academics in the university helped, very slowly, to make studying babies and children seem respectable.

Another important factor was technological. The video recorder was developmental psychology's telescope. Babies and young children don't communicate the way grown-ups do. The basic tools of grown-up psychology—multipart questionnaires, reaction-time button pressing, and the rest—are quite useless. And you can't dangle course credit or money or food pellets (the usual techniques with undergraduates and rats) in front of babies to get them to cooperate with you. In fact, two-year-olds will consistently do exactly the opposite of what you want them to do (in our lab a deep sigh goes up among the research assistants when we tell them it's time to test the twenty-four-month-olds).

Our basic tool is to observe babies' nonverbal behavior systematically. We watch their facial expressions, actions, and even eye movements. Unfortunately, anyone who has watched a baby knows that his or her behavior often seems formless and fluid, at least on the surface. You get a little glimpse of something, and then you wonder if maybe you just imagined it. This is part of what makes the Piagets' diaries so astonishing. But what makes a science really advance isn't just the astonishing geniuses, it's the methods that allow us ordinary idiots to do the same thing as the astonishing geniuses. By using videotape we can objectively measure what babies do and look at it slowly, over and over. We forget sometimes how recent this technology is. As late as the 1980s we were still hauling forty-pound reel-to-reel recorders to children's living rooms (as we often tell our students, at great length, when they start complaining).

It was a different technology, the digital computer, that gave developmental psychology a new theoretical justification. Neither Piaget nor Vygotsky quite had the theoretical tools to integrate the mind and the brain successfully. Success in science is often a matter of finding the right analogies, and the computer gave us a new one. Computers, after all, could do lots of things that seemed quintessentially mental, even quintessentially intelligent (like calculate equations or, the nerd's gold standard, play chess). And yet they were undeniably physical, material objects, just as brains are. The Big Idea, the conceptual breakthrough of the last thirty years of psychology, is that the brain is a kind of computer. That's the basis of the new field of cognitive science. Of course, we don't know just what kind of computer the brain is. Certainly it's very different from any of the actual computers we have now.

The new technology of video recording and the new theories of cognitive science turned the pioneering ideas of Piaget and Vygotsky into a full-fledged scientific research enterprise.

We could use camcorders to see children in a new way, and we could use the metaphor of computers to understand them in a new way. Most of all, we began to think that seeing and understanding children was something worth doing.

The payoff has been a set of startling scientific results. We've learned more in the last thirty years about what babies and young children know than we did in the preceding 2,500 years. We can explain everyday phenomena everyone knows about, like the terrible twos, and we've discovered exotic new phenomena that no one would have predicted, like the fact that newborn babies know what their tongues look like, that six-month-olds already can tell the difference between Swedish and English, and that two-year-olds know that some people actually like broccoli more than Goldfish crackers. And every new thing we learn about babies tells us something new about us; we are, after all, only babies that have been around for a while.

CHAPTER TWO

What Children Learn About People

The ancient problems of knowledge are all fascinating, but only the problem of Other Minds is gut-wrenching. We dedicate most of our waking life to deciphering the minds of others. Why did he do that? Is she telling the truth? Does she really not mind that I'm working late, or is she simply saying she doesn't mind on principle and then brooding about it? Does he really not know that I really *do* mind that he's working late but am saying I don't on principle? If a telepath were to monitor the great minds on the campus of any major research university, she would hear a lot more about sexual and office politics than about physics or chemistry (or, we confess, developmental psychology).

There are, of course, good evolutionary reasons for this. We are an intensely social species, deeply dependent on one another for our very survival. And we are also a complicated species, with a wider repertoire of actions than any other. We had to be able to predict what other people would do, even more than what woolly mammoths or saber-toothed tigers or flints and fires would do. One of the best ways of doing that is to know what's going on in people's minds.

The new research in developmental psychology tells us that quite literally from the moment we first see other people, we see them *as* people. To be a person is to have a mind as well as a body, an inside as well as an outside. To see someone as a person is to see a face, not a mask; a "thou," not an "it." We arrive in the world with a set of profound assumptions about how other people are like us and how we are like other people.

But the research also tells us that those innate assumptions are just the beginning, not the end, of our understanding of the mind. Identifying the essential personhood, the thouness, of all human beings may be enough for God and Martin Buber, but it apparently isn't for the rest of us sinners. We must also learn just what kind of a thou we're dealing with. Does he actually *like* broccoli? Will she go ballistic if I so much as touch that vase? When that kid on the playground said golf balls explode if you cut into them, was he lying or misinformed or dangerously crazy? These are the sorts of problems children face, and solve, as they get older.

Understanding the people around you is also part of becoming a particular sort of person yourself. As children learn what other minds are like, they also learn what their own minds are like. They learn how to have an ancient Greek mind, or a Dutch seventeenth-century mind, or a late-twentieth-century West Coast mind. (One of our children, just three years old, once suggested on a boring, rainy day that we should really go get a latte and check out some bookstores.) Communities have distinctive ways of thinking and feeling as well as dressing and eating, and children must learn these ways of being from the grown-ups around them.

We said in Chapter One that there are three elements in nature's solution to the problem of knowledge: innate knowledge, powerful learning abilities, and unconscious tuition from adults. All three of these elements play a role in the solution of the Other Minds problem.

What Newborns Know

You're lying in bed in the labor room of the hospital and you're about as exhausted, as utterly worn out, as you'll ever be. Giving birth is this peculiar combination of determination and compulsion. It's you pushing, and you push in a more concentrated, focused way than you've ever done anything, but in another sense you don't decide or try to push or even want to. You are just swept away by the action. It's like a cross between running a marathon and having the most enormous, shattering, irresistible orgasm of your life.

And then suddenly, in the midst of all this excitement and action, agitation and exhaustion, there is a small, warm body lying on your chest and a tranquil, quiet, wide-eyed face looking up at yours. Maybe it's just the natural endorphins flowing through you once the actual pain is gone, but instead of collapsing, as you might expect, you feel a kind of intensified alertness. You're preternaturally awake, and everything is clearer and sharper than usual. And through the next night or two, when the nurses have finally left you alone, and the helpful husband has gone home to get some sleep and tell the relatives, you lie with the baby in your arms and inhale that peculiar, sweet, animal, newborn smell, and you look, and look for an hour at a time, at the small, still somewhat squished face. And the baby, perhaps also under the spell of the endorphins, alert as he won't be for some days to come, looks at you. And then and there—before the sleepless nights and diapers and strollers and snowsuits have kicked in—that gaze seems to signify perfect mutual understanding, complete peace, absolute happiness.

That's the romance of it, anyway. The romance doesn't come, sadly, with every birth, just as the parallel romance of true love doesn't come with every sexual encounter. But, just as with true love, it is one of the great gifts of life and seems more than worth the risk of disappointment and the reality of pain.

How does this romance of instant understanding compare with the scientific reality? The recent research gives us a picture that is surprisingly in tune with the intuitions of new mothers. For many years "experts" who, in fact, knew nothing systematic about babies, took a certain perverse satisfaction in assuring parents that their new babies' minds were somewhat less sophisticated than that of the average garden slug. Babies couldn't really see; their smiles were "just gas"; the idea that they recognized familiar people was a fond maternal illusion. As this sentence is being written, a columnist in the day's paper talks about his son's new baby brother and comments pityingly on the child's conviction that the baby recognizes him, when, of course, babies can't tell people from dogs. It's as if there is a double layer of folk wisdom about babies. Nearly everyone who actually interacts with them immediately thinks babies have minds. And yet often there's an almost equally immediate cynicism, mixed with distorted echoes of medical legends. You hear new parents say things like "I could swear that she recognizes me, only I know that she can't."

But why should you believe us instead of those benighted experts who thought babies couldn't really see? How can we say we actually do know what babies think? With the help of videotape, scientists have developed ingenious experimental techniques to ask babies what they know. One whole set of techniques has been designed to answer two simple questions: Do babies think that two things are the same or different? And if they think they're different, do they prefer one to the other? You can present babies with pairs of carefully controlled events and see whether they can differentiate between them and which they prefer to look at or listen to. For instance, you can show babies a picture of a human face and a picture of a complicated object, like a checkerboard. Then an observer, who doesn't know what the babies are looking at, records their eye movements. By analyzing the babies' eye movements, you can see which picture they looked at longer. You can take the same idea a bit further by getting babies to suck on pacifiers that turn on different video- or audiotapes and determining which tapes they are willing to do some work for. You can see, for instance, if they will keep a tape of their own mother's voice playing longer than a tape of a stranger's voice.

Finally, you can exploit the fact that babies, like the rest of us, get bored. If you show babies the same old same old over and over, they stop looking and listening. Change the tape to something new, and they perk up and take notice. Developmental scientists call this boredom "habituation." So, for instance, you can show a series of different happy faces, and the babies will gradually lose interest; they'll habituate. Show them a new happy face, and they hardly look any longer. But show them a sad face, and they start to stare again. This means that babies somehow know that the happy faces are the same and the sad face is different.

Using these sorts of techniques we can show that at birth, babies can discriminate human faces and voices from other sights and sounds, and that they prefer them. Within a few days after they're born, they recognize familiar faces, voices, and even smells and prefer them to unfamiliar ones (it even looks as if they recognize their mother's voice at birth based on the muted but still audible sounds they hear in the womb). They'll turn toward a familiar face or voice and even toward a pad that has been held close to their mother's skin and turn away from other faces, voices, and smells.

Within the first nine months, before babies can walk or talk or even crawl, they can tell the difference between expressions of happiness and sadness and anger, and even can recognize that a happy-looking face, a face with a smile and crinkly eyes, goes with the chirp of a happy tone of voice. You can show them two films, side by side, one of a face with a happy expression and one of a face with a sad expression. If you turn on a sound track playing either a happy or sad voice, babies will look longer at the face displaying the emotional expression that matches the emotion they hear.

They even know how people move. You can tie small, bright lights to someone's elbows, knees, and shoulders and then film that person moving in the dark. In the resulting film, only the spots of moving light are visible. To an adult this pattern of lights is clearly human and can even convey emotions; it's like a kind of simple cartoon. It turns out that babies also are able to differentiate this abstract pattern of moving lights from patterns that aren't human, and they prefer it. They seem to be so tuned to other people that an abstractly human pattern of lights is riveting.

Even the limitations of babies' vision make them pay special attention to people. It's a myth that newborn babies can't see, but babies are very nearsighted by adult standards, and unlike adults, they have difficulty changing their focus to suit both near and far objects. What this means is that objects about a foot away are in sharp focus and objects nearer or farther are blurred. Of course, that's just the distance from a newborn's face to the face of the person who is holding him or her.

Babies seem designed to see the people who love them more clearly than anything else.

The newborn's world seems to be a bit like the room full of Rembrandt portraits in the National Gallery of Art in Washington, D.C. Brightly lit faces, full of every nuance of movement, life, expression, and emotion, leap out from the background of gloomy obscurity, in a startling psychological chiaroscuro.

All this, though, is just appearances. Do babies have a deeper conception of what it is to be human? There is some reason to think that they do. Twenty years ago one of us, Andy, made a startling discovery. One-month-old babies imitate facial expressions. If you stick your tongue out at a baby, the baby will stick his tongue out at you; open your mouth, and the baby will open hers. How do we know that this is really imitation, that we aren't just reading it into the babies' endlessly mobile faces? Andy systematically showed babies either someone sticking out his tongue or someone opening his mouth. He videotaped the babies' faces. Then he showed the tapes of the babies' faces to someone else, someone who had no idea which gesture the babies had seen. This second person had to say whether each baby was sticking out his or her tongue or opening his or her mouth. It turned out there was a systematic relation between what the babies did, judged by this necessarily neutral and objective observer, and what the babies saw.

At first Andy did these experiments with three-week-olds. But to demonstrate that this ability was really innate, he had to show that newborn babies could imitate. So he set up a lab next to the labor room in the local hospital and arranged with parents to call him when the baby was about to arrive. For a year he would wake up in the middle of the night, or dash out of a lab meeting, and rush to the hospital, in almost as much of a hurry as the expectant parents themselves. But that meant he could test babies less than a day old; the youngest baby was only forty-two minutes old. The newborns imitated, too.

At first glance this ability to imitate might seem curious and cute but not deeply significant. But if you think about it a minute, it is actually amazing. There are no mirrors in the womb: newborns have never seen their own face. So how could they know whether their tongue is inside or outside their mouth? There is another way of knowing what your face is like. As you read this, you probably have a good idea of your facial expression (we hope intense concentration leavened by the occasional smile). Try sticking out your tongue (in a suitably private setting). The way you know you've succeeded is through kinesthesia, your internal feeling of your own body.

In order to imitate, newborn babies must somehow understand the similarity between that internal feeling and the external face they see, a round shape with a long pink thing at the bottom moving back and forth. Newborn babies not only distinguish and prefer faces, they also seem to recognize that those faces are like their own face. They recognize that other people are "like me." There is nothing more personal, more part of you, than this internal sense you have of your own body, your expressions and movements, your aches and tickles. And yet from the time we're born, we seem to link this deeply personal self to the bodily movements of other people, movements we can only see and not feel. Nature ingeniously gives us a jump start on the Other Minds problem. We know, quite directly, that we are like other people and they are like us.

There are other reasons to think that even very young babies are especially tuned to people. Babies flirt. One of the great pleasures in life is to hold a three-month-old in your arms and talk absolute nonsense. "My, my, my," you hear your usually sane, responsible, professional voice saying, "you *are* a pretty bunny, *aren't* you, aren't you, aren't you, sweetums, aren't you a pretty bunny?" You raise your eyebrows and purse your mouth and make ridiculous faces. But the even more striking thing is that that tiny baby responds to your absurdities. He coos in response to your coo, he answers your smile with a smile of his own, he gestures in rhythm with the intonation of your voice. It's as if the two of you are engaged in an intricate dance, a kind of wordless conversation, a silly love song, pillow talk. It's sheer heaven.

But aside from being sheer heaven, it's also more evidence that babies spontaneously coordinate their own expressions, gestures, and voices with the expressions, gestures, and voices of other people. Flirting is largely a matter of timing. If you look around at a party, you can tell who's flirting just by looking at them, without even hearing a word. What you see is the way two people time their gestures so they're in sync with each other and with nobody else in the crowded room. She brushes her hair off her face, and he puts his hand in his pocket; she leans forward eagerly and talks, and he leans back sympathetically and listens. It's the same way with babies. When you talk, the baby is still; when you pause, the baby takes her turn and there's a burst of coos and waving fists and kicking legs. Like imitation, baby flirtation suggests that babies not only know people when they see them but also that they are connected to people in a special way. Like grown-up flirtation, baby flirtation bypasses language and establishes a more direct link between people.

The Really Eternal Triangle

So even in the first few months of life babies understand that there's something special about other people and that they are linked to other people in a special way. That's the Martin Buber part of our everyday understanding of the mind. But life isn't, unfortunately, all mystic communion or, even more unfortunately, just pillow talk. Even college students may resort to meaningless baby talk in the Valentine's Day classifieds, but they will not do very well if their term papers are equally vacuous. There is a lot more we need to know about people.

One thing to know about people is what they think about *things.* People look at things, want them, act on them, and know about them. When babies are about a year old, there is a striking change in the way they interact with people. Suddenly, instead of just you and me insulated in our cocoon of infant romance, interlopers enter the picture: teddy bears, balls, keys, rattles, lamp cords, spoons, puppy dogs, telephones, porcelain vases, lipsticks, distant airplanes—a panoply of fascinating, seductive, irresistible objects. As babies become able to sit up and reach and crawl, these things, which were formerly mostly the objects of a fascinated but distant gaze, become objects of desire and danger. Fortunately, other people don't completely vanish from the babies' thoughts (though it may feel that way to parents). Instead, they become an essential part of a kind of cognitive triangle.

When babies are around a year old, they begin to point to things and they begin to look at things that other people point to. Like imitation, pointing is something so familiar we take it for granted. But also like imitation, pointing implies a deep understanding of yourself and of other people. To point to something, especially when you point again and again, looking back at the other person's face until he or she also looks at the object, implies that you think, at some level, that the other person should look at the same thing you are looking at. We can systematically record and measure where babies look as they watch a grown-up point to particular places. By the time they're a year old, the babies will look, quite precisely, at just the place the grown-up pointed to.

Other experiments also show that one-year-olds have a radically new understanding of people. What happens when you show a baby something new, something a little strange, maybe wonderful, maybe dangerous—say, a walking toy robot? The baby looks over at Mom quizzically and checks her out. What does she think? Is there a reassuring smile or an expression of shocked horror? One-year-olds will modify their own reactions accordingly. If there's a smile, they'll crawl forward to investigate; if there's horror, they'll stop dead in their tracks.

Again we can show this quite systematically. For instance, a grown-up can look into two boxes. She looks into one box with an expression of joy and into the other with an expression of complete disgust. Then she pushes the boxes toward the baby, who has never seen inside the boxes. Nevertheless, the baby figures out something about what is inside just by looking at the experimenter's face: the baby happily reaches into the box that made her happy but won't open the box that disgusted her. The baby doesn't just understand that the other person feels happy or disgusted, but also understands that she feels happy about some things and disgusted about others.

In a similar way, one-year-old babies can figure out what to do with objects by looking at what other people do with them. Andy used imitation to test this. He would show babies a completely unexpected way to use a new object—he would touch his forehead to the top of a box, and it would light up. The babies watched in fascination, but they weren't allowed to touch the box themselves. A week later the babies came back to the lab. This time Andy just gave them the box, without doing anything to it himself. But the babies immediately touched *their* foreheads to the top of the box. There's a common myth that babies have no memory. But all during that week the new information about what people do with this thing had been percolating away. Moreover, the babies seemed to assume that if other people do something special to an object, they should do the same thing. (You can see this in the way babies play with a toy telephone. Even though the toy telephone doesn't actually do anything, babies will mimic what grown-ups do with a real telephone; they push the buttons and hold the toy phone up to their ear and even babble into it.)

When babies are around a year old, then, they seem to discover that their initial emotional rapport with other people extends to a set of joint attitudes toward the world. We see the same objects, do the same things with those objects, even feel the same way about those objects. This insight adds a whole new dimension to the babies' understanding of other minds. But it also adds a whole new dimension to babies' understanding of the world. One-year-old babies know that they will see something by looking where other people point; they know what they should do to something by watching what other people do; they know how they should feel about something by seeing how other people feel.

The babies can use other people to figure out the world. In a very simple way, these one-year-olds are already participating in a culture. They already can take advantage of the discoveries of previous generations. They don't have to discover for themselves that there is something worth looking at in that corner or something disgusting in that box, or that the box lights up when you touch it with your forehead. Even without using language we can tell them all those things. Even babies who can't talk yet are naturally cultural beings.

This new understanding also lets babies use other people to get things done. A one-year-old can point to a toy that's out of reach and expect that the grown-up will get it for her, or put her hand on the grown-up's hand and get the grown-up to spoon out the applesauce. Even before babies can talk, they can communicate.

This particular triangular story has a happy ending. The babies' new interest in things also leads to a deeper commonality and communication with other people. After all, there is more to communication than communion. Even with grown-ups, pillow talk eventually gives way to the rather different delight of discovering that you both really do love Thai food and hate Quentin Tarantino movies. In the best romances, you face the world together; you don't just face each other. After twelve months or so the same thing happens in the romance between children and their parents.

Peace and Conflict Studies

Babies and grown-ups, then, seem to work together to negotiate the perils of the other object successfully. But there is a more profoundly dangerous snake lurking in the infant Eden. The enemy within is always more potent than the enemy at the gates. As babies learn that people usually have the same attitudes toward objects as they do, they are setting themselves up to learn something else, something more disturbing: sometimes other people don't have the same attitudes they do. What happens when the baby reaches for the forbidden lamp cord, the porcelain vase, the lipstick? Or when her father is helping to direct repulsive mashed turnips, and not delicious applesauce, toward the baby's mouth? Commonality and communication fall apart.

It must seem positively paradoxical, even perverse, to the one-year-old baby. The more clearly he indicates his passionate desire for the lamp cord, the more adamantly his mother acts to keep it away. The more plainly she refuses the turnip, the more determinedly her father presents it to her. Even though both the baby and the grown-up are reacting to the same object, their attitudes toward the object seem to be different, even diametrically opposed.

By the time babies are about one-and-a-half years old, they start to understand the nature of these differences between people and to be fascinated by them. Again we can demonstrate this systematically. Alison and one of her students, Betty Repacholi, showed babies two bowls of food, one full of delicious Goldfish crackers and one full of raw broccoli. All the babies, even in Berkeley, preferred the crackers. Then Betty tasted each bowl of food. She made a delighted face and said, "Yum," to one food and made a disgusted face and said, "Yuck," to the other. Then she put both bowls of food near the babies, held out her hand, and said, "Could you give me some?"

When Betty indicated that she loved the crackers and hated the broccoli, the babies, of course, gave her the crackers. But what if she did the opposite and said that the broccoli was yummy and the crackers were yucky? This presented the babies with one of those cases where our attitude toward the object is different from theirs, where we want one thing and they want something else. Fourteen-month-olds, still with their innocent assumption that we all want the same thing, give us the crackers. But the wiser (though, as we will see, sadder) eighteen-month-olds give us the broccoli, even though they themselves despise it. These tiny children, barely able to talk, have already learned an extremely important thing about people. They've learned that people have desires and that those desires may be different and may even conflict.

We can demonstrate this discovery in the laboratory, but it is also dramatically apparent in ordinary life. Parents all know, and dread, the notorious "terrible twos," when the adorable if somewhat out-of-hand one-year-old rogue becomes a steely-eyed two-year-old monster out of melodrama. What makes the terrible twos so terrible is not that the babies do things you don't want them to do—one-year-olds are plenty good at that—but that they do things *because* you don't want them to. While one-year-olds seem irresistibly seduced by the charms of forbidden objects (the lamp cord made me do it), the two-year-olds are deliberately perverse, what the British call bloody-minded. A two-year-old doesn't even look at the lamp cord. Instead his hand goes out to touch it as he looks, steadily, gravely, and with great deliberation, at you.

Children may add various demonic variations. According to her mother, the most perverse of the authors of this book used to add further insult by holding out her hand to be slapped at the same time. On the other hand, one of our most charming and conciliatory children would produce his most radiant smile as he moved toward the forbidden object. Any hint of an answering smile was fatal. Another of our children would move closer and closer to the forbidden object in geometrically precise increments until she was only millimeters away from it, staring at her father all the time.

But this perverse behavior actually turns out to be quite rational. Just as experiments with very young babies explain our parental intuition that we have a special kind of rapport with our newborns, experiments with toddlers explain our intuition that that rapport sometimes breaks down when they get older. Two-year-olds have just begun to realize that people have different desires. Our broccoli experiment shows that children only begin to understand differences in desires when they are about eighteen months old. Fourteen-month-olds seem to think that their desires and ours will be the same. The terrible twos seem to involve a systematic exploration of that idea, almost a kind of experimental research program. Toddlers are systematically testing the dimensions on which their desires and the desires of others may be in conflict. The grave look is directed at you because you and your reaction, rather than the lamp cord itself, are the really interesting thing. If the child is a budding psychologist, we parents are the laboratory rats.

It may be some comfort to know that these toddlers don't really want to drive us crazy, they just want to understand how we work. The tears that follow the blowup at the end of a terrible-twos confrontation are genuine. The terrible twos reflects a genuine clash between children's need to understand other people and their need to live happily with them. Experimenting with conflict may be necessary if you want to understand what people will do, but it's also dangerous. The terrible twos show how powerful and deep-seated the learning drive is in these young children. With these two-year-olds, as with scientists, finding the truth is more than a profession—it's a passion. And, as with scientists, that passion may sometimes make them sacrifice domestic happiness.

There is also a more positive side to the two-year-old's new discoveries about people. One day, Alison came home from the lab in a state of despair that will be familiar to working parents. She had realized she was a terrible researcher (one of her papers had been rejected by a journal) and a failed teacher (a student had argued about a grade), and she came home to discover she was also a disgraceful mother (the chicken legs for dinner were still frozen). Like any good, strong, tough-minded professional woman in the same position, she broke down in tears on the sofa. Her son, who was not quite two, looked concerned and after a moment's thought ran to the bathroom. He returned with a large box of Band-Aids, which he proceeded to put on her at random, all over; this was clearly a multiple–Band-Aid injury. Like many therapists, he made the wrong diagnosis but his treatment was highly effective. She stopped crying.

This isn't just a touching story about a particularly adorable child (though, of course, Alison does tend to tell it that way). Systematic studies indicate that two-year-olds begin to show genuine empathy toward other people for the first time. Even younger babies will become upset in response to the distress of others (we all know the disturbing way the baby will suddenly begin to howl when a marital argument starts). But only two-year-olds provide comfort. They don't just feel your pain, they try to allay it. The two-year-old monster is also the two-year-old ministering angel.

This kind of empathy demands the same sophisticated understanding of other people that we see in the terrible twos. To be genuinely empathic, you have to understand how other people feel and know how to make them feel better, even when you don't feel that way yourself. You have to know that the other person needs some Band-Aids, even if you don't—just as you know that the other person wants broccoli, though you don't, or that she wants you to stay away from that lamp cord that seems so desirable to you. Real empathy isn't just about knowing that other people feel the same way you do; it's about knowing that they don't feel the same way and caring anyway. Babies aren't born with this deep moral insight, but by the time they are two, they already have begun to understand it.

Changing Your Point of View

At the same time that children are learning about the differences between their desires and the desires of others, they are also learning about differences between what they can see and what other people can see. When one-year-olds point and follow the points of other people, they seem to have discovered that they can see the same thing another person sees. But, just as with desires, the corollary of that discovery is that there may also be differences between what I can see and what you can see. If you are in the other room or on the other end of the phone, for instance, it will be quite useless for me to show you my latest artistic creation. This fact, though, seems lost on very young children.

We can test children's understanding of this difference by giving them hiding-and-finding games, a toddler passion. I can hide something from you and you can hide it from me. Very young toddlers love hide-and-seek, but they aren't very good at it; a two-year-old's favorite hiding stratagem may be to stick her head under a table with her overalled bottom in full view.

We can show this more systematically. For example, you can give children a long tube with a picture at the end of it, designed so that only one person can see the picture at a time. Then you can ask them to show the picture to Dad. Very young children will swivel the tube back and forth between themselves and Dad as if they can't quite imagine that Dad can see it when they don't.

Alison and Andy designed an experiment to test this idea further. First they set up an imitation game: you give the toy to me and I'll give it to you; you put the sticker on my hand and I'll put it on your hand. Children are very good at this and love doing it. Then Alison and Andy put a screen on the table between the experimenter and the child. The experimenter hid a toy from the child by placing it on her side of the screen. Then she gave the toy to the child and asked him to hide it from her. To do this correctly, the child had to put the toy on *his* side of the screen so that he could see it and the experimenter couldn't. But the youngest children, twenty-four- and thirty-month-olds, would often put the toy on the experimenter's side of the screen so that it was hidden from their own sight but was completely visible to the other person. And the toddlers actively experimented with this problem. They would walk over to the experimenter's side of the table to see how the screen looked from that side. Or they would invent ingenious ways of avoiding the problem, like hiding the toy behind their back so that it was hidden from everyone. Just as with the picture at the end of the tube, they couldn't seem to get their minds around the idea that they could see the toy but someone else couldn't.

Before they are three, though, children do learn about the differences between what they see and what other people see. A thirty-six-month-old, barely turned three, will always hide the toy correctly on *his* side of the screen. He knows that the other person can't see it even though he can himself. He can predict quite explicitly when you will see the object and he won't; he'll tell you that you can't see it but he can. Three-year-olds can even tell you about what an object looks like from different perspectives. If you put a yellow toy duck behind a piece of blue plastic, it will look green. You can show this trick to three-year-olds and let them see that the duck really is yellow. Three-year-olds will say that the duck looks green to the person on one side of the plastic but looks yellow to the person on the other side. Contrary to much conventional wisdom, these very young children are already beginning to go beyond an egocentric understanding of other people.

It is somewhat astonishing to see the very young two-year-olds make mistakes in solving such obvious problems. But it is even more astonishing that in a mere three years children learn something as important as the fact that people can, literally, see things differently. Of course, children still have a lot to learn about how different people understand the world differently. Apparently, grown-ups have a lot to learn, too; hence all the books about the gulfs between men and women. But we take the first steps toward that understanding soon after we take our first steps.

The Conversational Attic

If you call up the right Internet site, you will find yourself wandering through an enormous database called CHILDES—a database that contains literally millions of examples of young children's spontaneous conversation, recorded by linguists over the last thirty years. It's like finding a trapdoor in the ceiling that leads to an attic full of worn-out overalls and broken toys and faded snapshots. The CHILDES archives have some of that same poignancy, that sense of being surrounded by faint, ghostly voices. Abe and Sarah and Ben and Nina, children who have long ago grown up, had children of their own, even died, are eternally two years old here, nervously wondering about Dracula's true nature, seeing if eggcups can be eyeglasses, patiently explaining things to their parents.

In addition to letting us revisit these landscapes of lost childhood, CHILDES lets us look systematically at what very young children say and how the kinds of things they say change as they grow older. Among the very first things children talk about are desires, perceptions, and emotions. *Want* and the terribly-two *no; see* and *all gone; happy* and *sad* are among their earliest words. As we might expect from the experiments, by the time they are three, children also talk spontaneously, even ruminatively, about differences in desires, perceptions, and emotions. One of our own children discovered sadly that the much-anticipated dessert of a Sunday dinner, pineapple in kirsch, did not suit the two-year-old palate. For weeks afterward, apropos of nothing, he would suddenly say, "You know what, Mommy, pineapple is yummy for you, but it's yucky for me."

CHILDES reports a similar exchange between a New England three-year-old and his mom, contemplating the notorious exoticism of French cuisine. Child: "Can you eat snails?" Mom: "Some people eat snails, yes." Child: "Why?" Mom: "Because they like them." Child: "Mommy, do you want to eat snails?" Mom: "No, I don't think I'd like to eat snails." Child: "I don't like to eat snails [long pause] . . . people eat snails."

Another of our children, going to a *Star Wars* movie for his fourth birthday, produced the following bit of logical argumentation as the theater darkened and giant storm troopers filled the screen: "Four-years-olds don't be frightened. I'm a four-years-old today. I'm not frightened." One of the CHILDES children, Nina, puts eggcups on her eyes and then on her dad's eyes, and then takes them off and puts them on again, repeatedly saying, "Can see it, can't see it," as she does, commenting on the differences between what she sees and what her dad sees.

As soon as they can talk, children talk about both their own minds and the minds of other people. At first, though, they focus on desires, perceptions, and emotions: what they and others want, see, and feel, rather than what they know or think about. They only begin to talk about thoughts and beliefs later on. They begin to say things like "The people thought Dracula was mean, but he was nice" or "It's a bus, I thought it was a taxi."

Learning About "About"

Desire, perception, and emotion are central aspects of the mind, and it's remarkable that young children understand these mental states so well by the time they are barely three. All three of these states, though, have a quality that philosophers call transparency. Children seem to think about desires, perceptions, and emotions almost the way we might think about magnets or bullets. If you want something, you'll be drawn to it, just as filings are drawn to a magnet. If you're pointed toward an object, you'll see it, just as you'll be hit if you stand in the way of a speeding projectile. That's what philosophers mean by transparency.

But other mental states—beliefs and thoughts, for instance—are rather different. We can't just believe or think an ice cream cone or a toy car the way we can see or want one. Instead, we believe or think things *about* the ice cream cone or the car. We believe that the ice cream has nuts in it or that the toy car needs batteries. Philosophers say that belief is "opaque," or "representational." The classical philosophical example, back from the days when novels were published anonymously, was that you might not have the same beliefs about "Sir Walter Scott" that you did about "the author of *Waverley*," even though, in fact, "Sir Walter Scott" and "the author of *Waverley*" were one and the same person. When we say that someone believes something, we mean that that person has a kind of internal description or picture that refers to the thing under consideration, that is "about" it.

One important consequence of this is that beliefs can be false. We might think that the ice cream has nuts in it when really it's plain vanilla, or that the car needs batteries when it actually runs on a rubber band, or that Sir Walter Scott *isn't* the author of *Waverley*. We can believe something about the world that isn't true. Desires and perceptions can't be false in quite the same way.

How can we tell if children understand beliefs in this opaque, representational way? We can give them a simple scenario in which someone believes something that isn't true. For instance, we can show the children a familiar candy box. Anyone who sees it will leap to the conclusion that there's candy inside. When we open it, it turns out to be a trick: there are actually pencils inside. Then we can ask the children simple questions about this series of events. What did you think was inside it? What will your friend Nicky think is inside it, if he sees it all closed up like this? If someone wants candy, will he be surprised or disappointed when he looks in the box? If someone wanted pencils, would she look in the box? Does it seem as if the box has candy in it or pencils in it? And so on.

All these questions get at the same basic idea. The trick box leads us to have a false belief; it seems like one thing but really it's something else; it makes us represent the world in a way that's different from the way the world actually is. To adults, the answers to these questions seem obvious. But young three-year-olds consistently get these answers wrong. They say that everyone will know there are pencils in the box, that it seems as if there are pencils in the box, even that they always thought there were pencils in the box. It's as if the children think that since there is only one world out there, a single reality, everyone will understand it the same way. People will never have different beliefs about the same thing, and they themselves will never change their minds about anything.

Of course, adults also may share these illusions at times. But in us even the most stubborn dogmatism is tempered by the knowledge that at least sometimes people disagree with us and at least sometimes we turn out to be wrong. The children's doctrine of infallibility is completely unchallenged.

There is something especially odd about the fact that children don't seem to realize their own beliefs have changed. Alison decided to see whether children would make the same mistakes about their own beliefs that they made about the beliefs of others. The very first child she tested exclaimed, "Candy!" when she first saw the box, and then, "Holy moly! It's pencils!" when the box was opened. And yet barely a minute later she adamantly, and to all appearances sincerely, denied that she had thought there was candy in the box.

Alison has done other experiments that point in a similar direction. For example, three-year-olds seem to be unable to remember how they learned about something, even when the events took place only a few moments before. In one study the experimenter hid a cup under a cloth "tunnel," a wire arch covered with cloth, with an opening at either end. Children found out what was underneath the tunnel in one of three ways: they picked up the tunnel and saw the cup, they put their hands in the tunnel through the openings and felt it, or the experimenter simply told them, "There's a cup inside." Then she asked the children what was under the tunnel. They always got that answer right. But the next question was harder. She asked, "How do you know there's a cup in the tunnel? Did you feel it, or did you see it, or did we tell you about it?" Children were confused about how they had found out about the object. They said, for example, that they had seen the cup when actually she had told them about it. (These experiments have obvious implications for very young children's eyewitness testimony. Children aren't any more likely to lie than adults, and they don't confuse fantasy and reality, but they may well confuse what they saw and what a well-meaning lawyer or social worker told them.)

And yet children's memory about everyday events is excellent, as good as or even better than adults'. Recall that even the one-year-olds in Andy's imitation study could remember for a week that the experimenter had touched his forehead to the box. In other work he demonstrated that eighteen-month-olds remembered this novel event for as long as four months. And most of us are readily whupped by little kids at card games like concentration. Moreover, in our experiments we showed systematically that the children could remember past events, even though they couldn't remember their earlier beliefs.

What's going on here? These children challenge one of the oldest and most cherished philosophical doctrines, a doctrine that is sometimes called first-person authority. It seems to most of us that the Other Minds problem really is about others. While we must infer the thoughts of other people, we at least know for certain what we think ourselves. In fact, Descartes argued that the *only* thing we really know for certain is what we think ourselves; "I think, therefore I am." The children, though, make just the same mistakes whether they are reporting their own mental state or predicting the mental states of other people. It's as if they have a single theory about the mind, which they apply both to themselves and to others. They don't seem to understand their own minds any better than they understand the minds of the people around them. It may seem that we learn about other people by comparing them with ourselves. But, in fact, the research suggests that we also learn about our own minds by observing other people.

The Three-Year-Old Opera: Love and Deception

It may seem that the scientist-child we have been describing so far is a detached, distant observer. But finding out about other people profoundly shapes the way we feel about them and live with them. Finding out about ourselves profoundly shapes the way we feel and live, period. We saw how the eighteen-month-olds' new discoveries influence their everyday lives, as they become terrible, but also empathic, twos. The three- and four-year-olds' increasingly sophisticated understanding of the mind also influences their everyday lives.

Young children are certainly intensely emotional creatures. But the prevailing theories about children have treated these emotions as if they were the mainsprings of children's actions, unmodified by thought or knowledge. The Freudian tradition treated children's thoughts and beliefs as if they were entirely shaped by their primitive drives. Another, more biological tradition was inspired by work on the "bonding" of mother animals and their babies. Some baby birds, for example, follow after the first large moving thing they see after they are born (the comparative psychologist Konrad Lorenz, who studied this phenomenon, used to stroll through the streets of his Austrian village with a train of goslings behind him).

At least in some theories, human "bonding," or "attachment," was similarly seen as an almost reflexive, instinctual response. There is a perhaps apocryphal story that one maternity ward had a sign on the wall reading "Please do not remove babies from mothers until after bonding has taken place." A not at all apocryphal story is that when Alison gave birth to one of her babies, a nurse arrived to say she would be taking the baby away to the nursery. When Alison politely declined to be separated from her newborn (actually, she said they would need a crowbar), the nurse replied, "Don't worry, dear, we'll let you bond first." This was a sort of superglue view of bonding: put mother and baby together and hold for several minutes until firmly attached.

The more recent research changes this picture. Lusts don't dictate thoughts, and "bonding" isn't a once-and-for-all event that must take place in a critical period. Knowledge guides emotion more than emotion distorts knowledge. The relations between parents and children, like other human relations, develop and change as both partners come to know and understand each other better.

The new idea about attachment, or bonding, is that babies and young children develop "internalized working models" that are systematic pictures of how people relate to one another—theories of love. Of course, these models are heavily influenced by children's observations of the people around them. And also, these models, like scientific theories, influence the ways children interpret new observations. If you see that the people you rely on for warmth and comfort turn away from you when you're in distress, that may influence your expectations about how other people will act and your interpretations of what they actually do. But rather than being fixed, these internalized working models are actually flexible. Like scientific theories they can be changed with enough new evidence. As children get new information about how people work, especially how people work together in intimate ways, they modify their own views. Even abused children often seem to escape long-lasting damage if there is somebody around who doesn't turn away. A relatively brief experience of a friend or an aunt or a teacher can provide children with an alternative picture of how love can work.

Other things that children learn about people may also deeply influence their emotional and social lives. One of the things that Freud got right was the startlingly erotic character of three-year-olds. (We're developmental psychologists, and it still startles us.) Three-year-olds do act like lovers toward their parents. In fact, they act like lovers out of Italian opera, with passionate and sensual embraces and equally passionate despair at separation and jealousy of rivals.

But these passions may reflect real discoveries. The interactions of infancy involve a kind of concord between babies and the people around them, that pillow-talk feeling of inseparable intimacy. As babies become toddlers and then preschoolers, they increasingly realize that other people are separate psychological beings—beings with other desires, other emotions, other thoughts and beliefs. But it's just that sense of "otherness" that is at the root of erotic emotion. As we begin to realize that the ones we love are different from us, with other wants and thoughts and even other loves, we can't take their love for granted in the same way.

Preschoolers in love with their parents are more like Proust's Swann in love with the enigmatic Odette than like Oedipus in love with Jocasta. They aren't just in the grip of a primitive fatal ignorance; instead, they are haunted by an equally fatal knowledge. Part of love is wanting things (undivided attention, complete devotion, utter loyalty) you know you can't get.

Children's discoveries about belief also have consequences for other aspects of their relations to people. To deceive people, or to recognize that they are deceiving you, you need to be able to understand the differences between what they believe and what you believe. Doing that depends on understanding the way beliefs work. It depends on knowing what you have to do to make someone believe something that isn't actually true. Two- and three-year-olds are such terrible liars, they hardly qualify as liars at all. A three-year-old will stand on the other side of the street and yell back to you that he *didn't* cross it by himself. They are terrible liars just because they don't seem to understand what it takes to make someone have a false belief. We can show systematically that "real" lies only begin to appear at about four, at the same time that children start to understand "false-belief" problems like the deceptive candy box. Similarly, children only begin to understand that they can be deceived at about that age.

While learning to lie may not, at first, seem like a terrifically desirable skill, some kinds of deception are essential to civilized life. Children don't even seem to understand the necessary lies we call politeness until about four or five years old. They are baffled by a scenario in which someone pretends pleasure at an unwelcome birthday gift or hides the pain of a skinned knee under a show of stoicism. The idea that you could feel one emotion and yet express another seems contradictory to them. This may be why, in their everyday life, young children also have such a hard time masking their own emotions, another reason that life with a three-year-old can be like a twelve-hour-a-day performance of *Tosca.*

Knowing You Didn't Know: Education and Memory

Understanding beliefs in a new way also allows you to be instructed in a new way. We saw that young children don't seem to recognize that their ideas about the world have changed in the past or will change in the future. But this recognition, knowing that you don't know, is a prerequisite for the kind of systematic instruction that children get in school. In our culture, "preschool" ends and "real school" begins at about age six. In fact, even in cultures without "official" schooling, there has always been an implicit understanding that teaching three-year-olds is different from teaching six-year-olds. Throughout cultures and historical periods, only the older children have seemed suitable targets for formal instruction, in everything from the catechism to needlework to the protocols of knighthood. To take advantage of this sort of formal instruction you need to be able to do more than just learn and know (we've seen already that babies and young children are wizards at that). You have to know about knowledge and learn how learning works. You have to know what you need to learn and learn how to get new knowledge.

The fact that three-year-olds are so bad at recalling their own past beliefs may also tell us something important about memory. One of the great puzzles of psychology is the phenomenon of infantile amnesia, the fact that, as adults, we can't remember things that happened to us much before we were three years old. It's especially puzzling because, as we've seen, two- and three-year-olds and even infants seem to remember past events quite well. For adults this kind of continuous autobiographical memory depends on certain ideas about the mind. What makes our memories special is not just the fact that we know that certain things happened in the past, but that we know they happened to us. When we remember our past, we recapture not just what happened but what we thought and felt about what happened, how those events seemed to us. But, of course, that depends on being able to understand what it means to have thoughts. It depends on understanding how minds work.

Before we are three, our conception of our own mind, even our experience of our own mind, seems radically different from our adult conception and experience. When we are three, we don't seem to be able to understand the difference between our past thoughts and our present thoughts, though we do understand the difference between past events and present events. And we don't seem to recollect our past thoughts when they conflict with current ones. This may explain why we can't construct a continuous autobiographical story out of what happens to us.

How Do They Do It?

By the time we are five years old or so, then, we seem to understand the mind in much the way we will twenty or thirty years later. We understand thoughts and beliefs as well as we understand perceptions, emotions, desires, and feelings. Of course, completely understanding ourselves and other people is an arduous, lifelong business. But the basic groundwork is in place after only a few years. Even very young children, children who haven't yet learned to read or add two and two, *have* learned profound truths about their own mind and the minds of others.

The new developmental research has demonstrated consistent and apparently logical changes in children's understanding of people. We know less about what makes those changes happen. How is it possible for children to learn so much about the mind in such a short time? As we said in the first chapter, there seem to be three factors involved. Children can take advantage of an innately determined foundation, powerful learning abilities, and implicit tuition from other people.

Mind-Blindness

Autism is a disorder that affects about ten in ten thousand children. Most children with autism also have other difficulties; many are severely retarded. But some children with autism have normal intelligence. And yet there is something about the way they understand other people that makes them very different from the rest of us. Ask a bright twelve-year-old with autism what "proud" is and if he's ever been proud. There's a long silence. Finally, with furrowed brows, he mutters to himself, "I know that one." Then, hesitantly, he says, "Proud is, like, when someone scores a goal at soccer? That's proud?" He gets the answer right, but he seems to do it very differently from most twelve-year-olds, who provide immediate, unself-conscious anecdotes of vainglory. Temple Grandin, a woman with autism who is also a successful and well-known professor of animal husbandry, says that she feels like an anthropologist on Mars. Her knowledge of other people is painfully cobbled together by carefully observing the regularities in their behavior. Most of us are born with the ability to link our own mind and those of others. People with autism seem to have to solve the Other Minds problem from scratch.

The lives of people with autism make us realize how important it is to be able to understand other minds. Most children prefer people to things from birth. Children with autism often seem to reverse this preference. They're completely absorbed by patterns of blocks, or even train schedules, while they avoid other people. In a way, this makes sense. Imagine how terrifying and disturbing the world would be if you really did see other people as alien bags of skin moving in random and unpredictable ways, rather than as people with minds.

These differences are also systematically reflected in the way that children with autism behave in the many experiments we just described. Because most people with autism are also mentally retarded, the basic technique of these studies is to compare children with autism both with normally developing children of the same mental age and with children who are retarded for other reasons, for example, children with Down syndrome. Children with autism have trouble imitating facial expressions, they don't point or follow points in the same way, and they don't understand false beliefs, like the incorrect assumption about the deceptive candy box, until much later than normally developing children, or even children with Down syndrome. The whole unfolding drama of understanding the mind is absent.

Children with autism don't seem to have the fundamental presupposition that they are like other people and other people are like them. This unquestioned first principle, this axiom of our everyday psychology, is, paradoxically, part of what allows most children to go on to discover all the differences between themselves and others.

When autism was first described, back in the psychoanalytic 1950s, some psychiatrists decided that the disorder was caused by "refrigerator mothers," mothers who were cold and unresponsive to their children. Mothers with college degrees were especially likely to make their children autistic. It is almost unbearable to think of what it must have been like for mothers who were already dealing with this sort of tragedy to be told that it was not only their fate but their fault. (From our present vantage point, as feminist cognitive psychologists at the turn of the millennium, it may seem that criticizing Freudian attitudes toward women in the fifties is like shooting fish in a barrel. This particular barracuda, though, still grins so evilly that it seems more than worth a shot or two.) It is quite clear now that autism has nothing to do with how parents treat their babies. It starts very early, there is a strong genetic component, and in some cases it may also be caused by damage to the brain before birth. We even have some ideas about what parts of the brain are involved. The senselessly cruel mother here is Mother Nature.

Becoming a Psychologist

The difficulties of children with autism suggest that there is an innate foundation for our understanding of the mind. On the other hand, the extended unfolding of different types of knowledge about the mind suggests that we have to build on that foundation. And it is striking that even some children with autism can get there in the end, though it is plainly infinitely harder when your beginning is so different.

What can we say about the processes that let us learn about the mind? What takes us from our early conviction that we are like the people around us to our full-blown understanding of the nuances of desire, perception, belief, and even existential angst? We think that children learn about the mind by being psychologists. They make predictions, they do experiments, they try to explain what they see, and they formulate new theories based on what they already know. This seems like the best way to account for the unfolding succession of ideas about the mind that we've just described. We can observe children experimenting with ideas about the mind in their everyday play, and thoughtfully trying to explain the odd behavior of other people in their everyday language.

But we also have some more direct evidence for the idea that children learn like scientists. Alison and Virginia Slaughter, one of her students, looked at three-year-old children who didn't yet fully understand belief—children who still said they had always thought that there were pencils in the candy box. Then, over the course of a few weeks, Virginia gave the children systematic evidence that their predictions were false. She told them firmly that they hadn't said pencils at all, they had said candies. When the children predicted that Nicky would think there were pencils in the box, she dragged Nicky in and asked him. Another group of children got very similar training about number problems—problems that had nothing to do with the children's understanding of the mind. At the end of the two weeks she asked the children a new question about false beliefs (about a set of soaps that looked like golf balls). The children who had received counterevidence to their mistaken ideas about the candy box did much better on questions about the golf-ball soap than the children who had learned about numbers. But she also asked the children new questions about all sorts of other aspects of belief, questions such as where beliefs come from and how appearances and realities differ. The children who had gotten the counterevidence not only did better on the questions about the trick objects, they also did better on lots of other questions about belief. By providing just the right kind of evidence at just the right time, we seem to have provoked a big, sweeping change, a sort of theoretical revolution, in the way these children thought about the mind.

We think that children learn about other people, and that they learn the same way scientists learn about the world. At first they may just ignore counterevidence that contradicts their theory. In fact, three-year-olds will tell you that they said there were pencils in the box when they first saw it and will even maintain that Nicky said there were pencils in the box, when he actually said just the opposite. Gradually, though, as enough different kinds of contrary evidence accumulate, it's no longer possible to just ignore or reinterpret the facts. When the new theory finally replaces the old one, there are far-reaching implications. The new theory doesn't just let us deal with the contrary evidence; it also lets us understand many other phenomena in a new way. And the new theory lets us create a whole set of new predictions about what will happen in the future.

When Little Brother Is Watching

In the experiment we just described, we played the role nature plays for scientists, dashing expectations and shaking up preconceptions but also giving hints about where the right answers lie. But then we're developmental psychologists. Do ordinary people play the same role in children's natural lives?

In one sense the answer has to be yes. Just by being themselves, people give us much of the information we need to understand them and to understand ourselves. There is reason to think, though, that one rather neglected set of people—big brothers and sisters—plays an especially important role in children's understanding of minds. Usually older siblings do better than younger siblings on things like IQ tests. But, consistently, younger siblings do better on tests of their understanding of the mind. They are likely to understand the trick box problem at a younger age than older siblings. And the more brothers and sisters children have, the better they do.

Parents egocentrically tend to think that they are the deciding factors in their children's lives. But for a two-year-old, an older brother or sister may actually be a more enthralling exemplar of human nature. In many cultures, in fact, older siblings do most of the child care after babies are weaned. Even in our culture parents notice the way a two-year-old will devotedly follow every movement of her four-year-old brother, and the way the four-year-old will even change his tone of voice as he talks to the baby.

We don't yet know exactly how older siblings teach babies about the mind. But there are two broad, related possibilities. Remember that much of what children learn involves the differences between their own mind and the minds of others; they largely take the similarities for granted. In fact, the assumption that we are like other people seems to be part of the early understanding of the mind. Parents tend to minimize the distance between their own mind and their babies' minds. They look for commonality and understanding instead of difference, and their lessons are primarily lessons about convergence. (The terrible twos are so awful precisely because they force us to confront our differences.) This is partly, of course, due to the selfless virtue of us parents, but it may also be, less righteously, because our influence on the babies' world is so omnipresent and inescapable. Siblings may provide a counterweight. They are much more likely to emphasize differences between what they want and what the baby wants, or to witheringly contrast their highly superior four-year-old knowledge and the baby's pitiful two-year-old ignorance.

The other possibility is that children are particularly motivated to try to understand their older siblings and make accurate predictions about them. Napoleon remarked that the valet always knows more about his master than the master does about his valet, and this may apply to little brothers and sisters, too. Parents treat babies and toddlers well pretty much no matter what happens. Getting big brothers and sisters to do what you want requires a lot more guile, cunning, and specialized expertise. Younger siblings may use a special version of our general human survival strategy: outwit the big guys. And, in fact, there is other evidence that younger siblings tend to be more charming and socially skillful, if less ambitious and domineering, than older siblings.

Scientific explanation always increases rather than diminishes our sense of wonder and awe. It is still wonderful and awesome that babies learn so much so quickly, even when we begin to understand how they do it. And while we may occasionally condemn nature for her mindless cruelties, we have much to be grateful for. Evolution seems automatically to grant most children a fundamental capacity for intimacy, a profound psychological curiosity, and plenty of kinfolk to be intimate with and curious about. What more could we ask?

CHAPTER THREE

What Children Learn About Things

We seem to exist in a world of three-dimensional objects moving through space in regular ways, objects that are outside us and would continue to be there even if we weren't looking at them. Some of these objects are people, some are animals, some are plants, and some are just stuff. Some of the objects are similar to one another, and some are different. And yet all of this banal, everyday experience is somehow based on tiny events on the edges of our bodies, photons bombarding our retinas, molecules of air vibrating at our eardrums, flickers of pressure on our fingertips. How do we bridge this apparently insuperable gap between our rich everyday experience of the world and the impoverished information of our senses?

While versions of the Other Minds problem still engage us every day, the External World problem is curiously invisible. We take our ability to perceive and understand objects for granted. It's the groundwork for everything we do. It's a philosophical purloined letter, so simple and in such plain view that we have a hard time seeing it.

One way of making these deep assumptions visible is to think about what happens in a magic show. All magicians, from the moonlighting student in the tuxedo at the children's party to David Copperfield, produce similar effects (the main differences are in the patter and the stage setting). The magician's stock-in-trade is to produce (or seem to produce) events that violate our usually unquestioned beliefs about how objects work. Our almost visceral experience of shock and wonder reflects the strength of those beliefs. Magicians make objects move from one point to another without traversing the space in between: the white rabbit was in the box, and now it's in the top hat. They make what looks like one object into two objects: the single silver ring becomes two rings as we watch. They make objects seem to influence each other from a distance: the magician waves his magic wand, and the box on the other side of the stage wiggles back and forth. They transform objects from one state to another: the water turns into orange juice. They even turn an inanimate object into a living one: the silk scarf becomes a dove.

Why are magicians so surprising and interesting? And why are we so sure that the rabbit really didn't disappear from the box and reappear in the hat? It must be because we are quite certain that things can't actually disappear, or turn into two objects, or be influenced by a magic wand, or change from one thing into another. In fact, we are so certain that, as we say, "we don't believe our eyes." Magic shows make us realize the complicated, abstract set of principles of everyday physics that we usually take for granted. Those beliefs are so deeply ingrained and so important to us that they even override what we actually see. We know that the rabbit can't actually just appear in the top hat. Even little kids "get" magic; the teenager in the tuxedo gets just as excited a response as David Copperfield in Las Vegas.

Nobody ever explicitly teaches us that the magician's tricks are impossible: no one tells us that the rabbit can't be in the hat. Yet it doesn't seem feasible that we could learn these principles directly from our sensory experience either, from the bombardment of photons and sound waves. In fact, as we've learned more and more about how our senses work, we've realized how complicated and tortuous the path is from the world to our brains.

Aristotle could still think that the true essence of objects sailed through our eyes and into our minds, and that that was how we saw the world. It was as if each time you glanced at a page of this book, you were inhaling a tiny, perfect piece of "bookness." But by the seventeenth century we had begun to understand that vision was the result of light that was reflected off objects and into our eyes. Scientists (like Bishop Berkeley and René Descartes) were making advances in optics, the science of light and vision, at the same time that philosophers (like Bishop Berkeley and René Descartes) puzzled over how optical phenomena could ever lead us to know what objects were really like. How could patterns of light bouncing off our retinas tell us that objects can't disappear into thin air or that space is three-dimensional?

The modern answer to the question is that we have a special kind of knowledge that enables us to translate the information at our senses into representations of objects. Our brain takes sensory information, the patterns of stimulation at our retinas and eardrums, and systematically transforms that information. It rearranges and changes it in a way somewhat similar to the way your word processor can rearrange and change the sequence of symbols you type (though, of course, the brain produces much more complicated rearrangements than the word processor). The outcome of this process is the coherent, complex network of beliefs that are so shockingly challenged by magicians.

The evolutionary trick is that these beliefs—that is, the representations we finally end up with after all these transformations and rearrangements—really do tell us about the world out there. The brain ends up with a picture that is actually closer to the real structure of the world than the raw sensory information it started out with. The world really is populated by objects moving in space in regular ways, and evolution ensures that we eventually understand this.

Just as it's important to infer the nature of other people's minds in order to survive, it's also important to infer the nature of the physical world. Determining that your neighbor is sexually interested may improve your reproductive success, but not if you can't figure out how to negotiate your way through the forest trees to get to his hut. Determining whether your neighbor's mate is likely to throw a rock at you is important for your survival, but so is dodging the rock.

So you, as a grown-up, know how to decode the chaotic, ever-shifting patterns of light, touch, sound, and smell that surround you into a book, a couch, a Mozart CD, and a convenient cup of coffee. But where does that knowledge come from?

The broad lines of the answer are similar to the answer to the Other Minds problem: we know a lot to begin with; we learn much more; other people unconsciously teach us. Some kinds of knowledge are there at the start. Even newborns seem to know that we live in a three-dimensional world and that something that looks curved will feel curved, too. But other kinds of knowledge emerge only gradually. Babies don't, at first, seem to understand how objects can be hidden by other objects. Just as babies have to learn things about people that we take for granted, they also have to learn a surprising amount about simple physical objects. Finally, some kinds of information about objects seem to be unconsciously conveyed by grown-ups. We unconsciously "teach" babies about objects even in the language we use to talk to them.

What Newborns Know

The Irresistible Allure of Stripes

Let's go back to the small, warm creature in the hospital room. That brand-new baby is already deeply connected to other people, but that's not all that goes on in the baby's world. Babies love human voices and faces more than anything, but they also love stripes and edges. Babies only a few days old will gaze with focused, cross-eyed intensity at the corner of the ceiling or a striped shopping bag while they ignore all the expensive toys, with their bright colors and soft prints, that Grandma brought along. We can show this systematically in the types of experiments we talked about in the last chapter. You can show babies different kinds of pictures and see where they look. Babies will turn toward complex patterns of high contrast and away from simple patterns with little contrast. Checkerboards and bull's-eyes appear to be at the peak of newborn aesthetic sensibility. In fact, manufacturers of baby toys have taken advantage of this research: the patterns on mobiles designed for very young babies are often taken straight from the pages of academic developmental psychology journals.

Why do babies love stripes? It turns out that this question is as important to cognitive scientists as it is to toy makers because it helps answer another question: How do we divide up the continuous visual image in front of us into separate things? When you look at the book in front of you, how do you know where to draw the line between the book and the background of couch fabric behind it, or the edge of the hand that is holding it? Although this might seem like a simple ability, in fact, the most sophisticated computer vision systems have a very hard time doing it.

Images such as stripes, where there is a sharp contrast between the brightness and texture of two surfaces, are important because they usually indicate where objects begin and end. If you hold this book up against a background, you'll see that the areas of greatest contrast in the image you see, the edges, correspond to the real boundaries of the book. Camouflage works by introducing edges to the inside of an object and obscuring the edges between an object and its background.

If you give young babies a complicated picture and record their eye movements as they look at the scene, you'll see them tracing the outside edges of objects. Newborns are already imposing order on what William James called the "blooming, buzzing confusion" of their senses. They're already organizing the world into a bunch of different things. Paying attention to edges is the best way of dividing a static picture into separate objects.

The Importance of Movement

But, of course, the baby's world isn't static. Even in the hospital room, things are constantly moving. And even newborns will follow a moving object with their eyes. (Alison's older sons found great amusement in "hypnotizing" their baby brother by slowly moving a toy back and forth in front of him.) Movement provides even better cues about where objects begin and end than do just edges alone. Imagine a baby looking at a Big Bird doll lying on a bunny quilt. The doll may have a number of different parts, each of which has specific edges—the head is visually separate from the body, which is separate from the feet. In the same way, each of the bunnies on the quilt also has its own separate edge. But if you pull the quilt out from under Big Bird, all the quilt's edges will move together, and they will move on a different path from all the parts of Big Bird. Psychologists, with uncharacteristic poetry, call this the principle of common fate. When things move together on the same path, they must be part of the same object.

You can demonstrate systematically that babies pay attention to this sort of information. If you show very young babies a video of a static Big Bird that then explodes into its separately defined parts, they won't be perturbed. Because all the parts had separate edges anyway, they may, for all the babies know, have been separate objects to begin with. But if you show them Big Bird moving first, so that they see that all the parts of the object move together, and then show them the exploding Big Bird, they'll look much longer and more attentively, as if they recognize that something is wrong. Seeing the parts move together, seeing their common fate, seems to tell the babies that this is just one object and that its parts are eternally joined together. So babies already have some principles they can use to impose order on a chaotic world.

Movement seems to be important for babies in other ways, too. Very young babies already know a surprising amount about how objects characteristically move. Young babies not only can follow the movements of an object in front of them, they seem to be able to predict how an object will move in the future. Suppose you show the babies an object following a particular trajectory—that is, moving in a particular path at a particular speed—say, a ball rolling on the table. Now the ball rolls behind a screen. They will look ahead to the far edge of the screen, to the place where the object ought to appear if it keeps moving at the same rate and on the same path. If the object does appear there, the babies are unperturbed and keep following the object. But if the object doesn't appear there, or if it appears at the wrong spot or too quickly or too slowly, they look intently at the edge of the screen for much longer. Sometimes, in fact, they look back to the other edge of the screen, or look farther ahead along the path the object should have taken. They seem able to predict where the object should be and when it should get there.

Seeing the World Through 3-D Glasses

Objects have edges and objects move, but another important thing to know about objects is that they are three-dimensional. One of the classic philosophical debates of the eighteenth and nineteenth centuries was about how we turn the two-dimensional, flat image projected onto our retinas into a three-dimensional world. With a little effort (it helps to close one eye and cup your hand around your other eye to make a frame), you can almost see the world around you as a flat picture, admittedly a very well painted one. It's not quite as unsettling as seeing your loved ones as bags of skin, but it is pretty weird and very different from your ordinary experience. But that flat picture is what the image that reaches your eye actually is like.

The great British philosopher Bishop Berkeley argued that we had to learn that space was three-dimensional by coordinating our visual experience and our tactile experience of moving through the world. Berkeley thought that touch was the only sense that gave us direct information about distance and solidity; somehow that information had to be associated with the two-dimensional information we got from vision. Babies demonstrate that Berkeley was wrong.

For one thing, even tiny babies who can't yet walk or crawl act in ways that indicate they understand distance. If you show babies a "looming" ball—a ball that looks as if it's rapidly approaching them—the babies will shrink back and even put their hands protectively in front of them. In much the same way, if you show babies a seductively interesting toy within arm's reach, they'll extend their arms clumsily toward it, even though they're far too little to grab it successfully. When they're a bit older, they'll reach toward a toy that is within reach, but not toward a toy that's out of reach.

Even very young babies have what's called size constancy. Suppose we show you a ball and then show you the same ball twice as far away. The new image on your retina will be only half as large, but you'll have no trouble identifying it as the same ball. On the other hand, if we now show you a ball that's twice as far away and also twice as big, you'll think it's a new ball, even though this time the image on your retina is the same size it was at first. It's as if you implicitly calculate that objects that are farther away look smaller.

Young babies seem to do this as well. Remember that babies perk up and look longer when they see something new and stop looking if it's the same old thing. Suppose you show them the first sequence of events: the close ball followed by the faraway ball. Even though the image on their retinas has changed, they show no particular interest; they act as if the faraway ball is more of the same. However, they do perk up and pay attention when they see the big ball from far away, despite the fact that this time the size of the image on their retinas did not change. Babies, like us, seem able to go beyond the image on their retinas and perceive something about the real ball out there.

Another great English philosopher, John Locke, posed another classical epistemological problem. What would happen if you miraculously restored the sight of someone who had been blind from birth? Would that person recognize all the objects he had known so intimately through touch, or would he have to painstakingly learn that the smooth, hard, curved surface looked like a porcelain teacup, or that the familiar, soft, yielding swells and silky hairs translated into a visual wife? Locke thought that the blind man would have to learn to make connections between the two types of experience.

Babies are a more common miracle than suddenly cured blind men, and it turns out you can ask them Locke's question, too. They think Locke, like Berkeley, got it wrong. Andy gave one-month-old babies one of two pacifiers to suck on, either a bumpy one or a smooth one. The babies never saw the pacifiers. They just felt them. Then he let the babies look at bumpy and smooth objects, without letting them feel them. The babies looked longer at the object that was the same shape as the one they had just been sucking on. Somehow, they could relate the feel of the pacifier in their mouths with its visual image.

You can ask the same question about the relationship between sound and vision. Even newborns will turn their heads and look toward an interesting noise, suggesting that they already expect to see something in the direction of the noise. You can do more systematic experiments to test this, too. For instance, you can show babies two objects bouncing at different times and play an audiotape of a *boing, boing, boing* sound that is synchronous with only one of them. Babies can tell which visual display matches what they hear; they look longer at the one that bounces in sync with the audiotape.

Even more startling, Andy and Pat showed babies a silent video of a face saying either *ahhh* or *eeee,* and then they played the babies audiotapes of each vowel sound. Five-month-olds could tell which face went with which sound. They looked at the face with the wide-open mouth when they heard the *ahhh* sound and at the face with pulled-back lips when they heard the *eeee* sound. Babies evidently have a primitive ability to lip-read, at least for simple vowels. (This was a provocative experiment—all those wide-open mouths and *ahhhs*. Soon after they finished doing the study together, Andy and Pat got married.)

So in the first few months of life, babies already seem to have solved a number of deep philosophical conundrums. They know how to use edges and patterns of movement to segregate the world into separate objects. They know something about how those objects characteristically move. They know that those objects are part of a three-dimensional space. And they know the relationship between information that comes from their different senses—they can link the feel of a nipple and its pink protuberance, the sound of a voice and the moving lips they see, the ball's exuberant bounce and its accompanying *boing*.

The Tree in the Quad and the Keys in the Washcloth

If babies know all this, what's left to learn? Quite a lot, it turns out. One classic epistemological dilemma is how we know that objects are still there even when we don't see them. As you read this, only a small portion of the room is actually in view. The book itself is probably occupying most of your vision. But you have no doubt that you're in a room full of other things even if you can't see them right at this moment. A pair of philosophical limericks capture the problem:

> There was a young man who said, God
> Must find it exceedingly odd
> To think that this tree
> Continues to be
> When there's no one about in the quad.

One philosopher's answer was:

Dear Sir, Your astonishment's odd,
I am always about in the quad,
And that's why the tree
Continues to be,
Since observed by, yours faithfully, God.

Modern cognitive scientists can't be satisfied with this answer, but we can still ask the question. How do we come to be so sure that the tree is still there when we aren't looking at it?

We've noted that babies can already make some predictions about where an object will reappear once it vanishes from sight. For instance, if you show babies a rolling ball that disappears at one end of a screen, they predict that it will reappear at the far edge of the screen at the right time. To do this, babies have to be able to think about the object even when they can't actually see it. Another way of testing this capacity is to show them a sort of magician's trick. Suppose you show young babies the object disappearing behind the screen, and the object fails to reappear or shows up in an odd location. Babies in the first six months of life look at this sort of scene for much longer than a scene in which the object reappears where it should.

In other circumstances, though, babies act as if they don't know very much about what has happened to objects that disappear. Suppose you show a six-month-old some wonderfully fascinating thing, a watch or a bunch of keys. He grins and bounces excitedly and starts to reach for it. Now cover the keys with a washcloth. The baby stops dead in his tracks. The excited glee is replaced by a sort of blank puzzlement. Pull off the washcloth and the glee returns.

Piaget tried this first, and he articulated the puzzle: If the baby wants the keys so much, why doesn't he just pull off the washcloth and find them? Maybe it's because he just isn't well coordinated enough. But you can test that by covering the keys with a transparent cover. The baby has no trouble whisking off the cover then. Maybe the baby doesn't remember that the keys are there. But you can show that babies this age remember other events for days or weeks. Instead, it seems the baby genuinely doesn't think that the keys are still there under the washcloth. For him the keys' reappearance from under the washcloth is like the magician's rabbit in the hat for us, a mystifying act of legerdemain.

In fact, babies only gradually learn about hidden objects. By the time they're around nine months old, they can easily find the keys under the cloth, but there are other, more complicated hiding games they still will not understand. Suppose you take a fifteen-month-old and you show her the following sequence of events. You put the keys in your hand and carefully close your hand around them. Then you put your hand under a cloth and leave the keys there. Finally, you take your hand out and show her that your hand is now empty. To us, the conclusion seems obvious—the keys must be under the cloth. Surprisingly, though, the same baby who can confidently find the keys when you simply cover them with a washcloth is, once again, a picture of stupefaction and puzzlement. Babies don't exactly have jaws, more just extended cheeks, but if they did, this baby's jaw would drop. She searches your empty hand, turning it over and over, as if those darned keys must be there somewhere. She looks on the floor. She shrugs, makes an empty-handed gesture, and even says "Where?" or "All gone." She has no idea where the keys could be if they are not in your hand, where she saw them disappear.

It isn't as simple as saying that for the younger baby "out of sight is out of mind." Even the youngest babies can keep some aspects of objects in mind when they're out of sight. But it does seem that the young baby's conception of what happens during disappearances is very different from our grown-up conception. And that means the baby lives in a universe that is profoundly different from our own. For us, it seems absolutely obvious that the keys must be under the cloth no matter how they're put there—where else could they be? But this is not only not obvious to the baby; it's something that has to be painstakingly learned. The baby, at first, lives in a perpetual magic show, where objects often seem to whirl about from one place to another with no rhyme or reason. Figuring out how it's all really done is one of the most important and difficult intellectual challenges of infancy.

Making Things Happen

Yet another great eighteenth-century philosopher, David Hume, posed a classical philosophical puzzle. When we see that one event always follows another event, we're likely to conclude that the first event caused the second. If every time you drink a cappuccino after dinner, you find yourself locked in a 3:00 A.M. crisis of existential dread, you may eventually work out that it is the coffee, and not the fundamental meaninglessness of the cosmos, that is responsible. We draw these sorts of causal conclusions all the time, and they play an absolutely essential role in what we actually do. You may very well switch to drinking green tea if you would prefer to cultivate a Zen-like serenity, or, for that matter, take up late-night espresso if you want to avoid suburban complacency. But Hume pointed out that we have no intrinsic reason for thinking that one event caused the other, just as we have no intrinsic reason for thinking that other people have minds, that space is three-dimensional, or that sounds and sight are linked. We never actually see one event make another happen. All we see is that the occurrence of one event is consistently followed by the occurrence of another. Why do we conclude that one event causes the other?

It turns out that even very young babies make some assumptions about causal connections between events. Three-month-olds already know about one very important type of cause and effect: they know that their own actions can influence events in the world. In fact, in some ways the causal power of our own actions is the primordial type of causation. Perhaps that's why we are all convinced we have free will. It's as if we think of our own decision to act as the most basic kind of cause, that we are ourselves the real prime movers.

You can give even a tiny, helpless baby artificially enhanced causal powers. You simply tie one end of a ribbon to her foot and the other end to a mobile. When the baby kicks, the mobile moves. Even very young babies rapidly learn to kick the foot with the ribbon to make the mobile turn. If you present them with the same mobile a week later, they will immediately start kicking the appropriate foot. They won't kick if you show them a new mobile. So babies make some assumptions about how their actions will influence the world. Just as important, those assumptions allow them to learn genuinely new things about how the world works.

Some of the very young babies' assumptions about causality, however, seem pretty strange. Suppose you disconnect the ribbon from the mobile, right in front of the babies' eyes. Three-month-olds will go right on kicking, as if they expect their actions will do the trick all by themselves. Moreover, even when the babies are connected by the ribbon, they not only kick but also smile and coo at the mobile, as if they think winsome charm is just as likely to be effective as crude direct action. The babies seem to understand that doing something can make other things happen, but they don't yet understand that this needs to be done through intermediary physical processes. They don't seem to appreciate, for example, that they must be in direct contact with the object in order to make it move.

Piaget called actions like kicking your foot even when the ribbon is disconnected magical procedures, and they do seem to have a sort of superstitious quality. At the same time, though, you could also argue that the babies' behavior is perfectly rational, given the babies' experience. Very young babies may simply be mixing up two different types of causal processes, the kind that influence things (like kicking) and the kind that influence other people (like smiling and cooing).

As scientists we think that everything is mediated by physical causality of some sort, including our interactions with other people. There are, in fact, light and sound waves that go from one person to another even if we can't see them with the naked eye. But from our everyday point of view, it appears we are able to influence people without any direct physical contact at all. (It's probably that fact that makes telepathy seem plausible to so many people.) After all, just looking at someone across a crowded room can set quite a dramatic chain of events in motion. We influence people psychologically by communicating, talking, gesturing, and making faces—we don't have to touch them. In fact, trying to physically manipulate other people to get them to do what we want is usually quite counterproductive, if not actually illegal. Psychological causality is often our most powerful tool.

Psychological causality is particularly important for babies, not only because they can't push things around as much as we can, but because they have to get other people to satisfy most of their needs. When very young babies first try to influence the external world, they may not differentiate between physical and psychological causality, and this may lead to the apparently magical and irrational quality of many of their actions. They make the mistake of using psychological means to try to influence the physical world. Smiling and cooing can get a reaction from Mom even though you're not physically attached to her. It's as if they think maybe they'll have the same effect on the mobile.

In fact, much of what we think of as magical, irrational thinking in adult life may really reflect the same sort of confusion between physical and psychological causality. Shamans and magicians say special words, wave their hands in particular ways, and take care in choosing particular garments in order to influence events in their world. This may seem odd and irrational, but when you think about it, all of us do this when we're trying to influence other people (well, two out of three of us for the garments). If you can use words to get someone into a white-hot rage or into bed with you, why not try to use words to give someone a disease or make her pregnant? "Magical procedures" of this type, whether in children or in adults, are, in fact, ineffective, but believing in them may not really be irrational—just mistaken. They may be based on a confusion about where psychological causality leaves off and ordinary physical causality begins.

By the time babies are about a year old, there seems to be an important change in their understanding of causes. They seem to have learned something about the differences between psychological and physical causality, and they understand more about how physical causation works. They also know something about how events or objects can influence each other. Younger babies can learn to produce an action that has an effect in the world. For example, they can pull a cloth that has a toy on top of it toward them. The peculiarities and limitations of that understanding become clear, though, when you present the babies with a new, slightly different problem by putting the toy to one side of the cloth. The babies pull on the cloth just as intently and are startled to see that nothing happens, just as they keep kicking even when the ribbon is disconnected. By the end of the first year, though, babies no longer make this mistake; they seem to know right away that the object has to be on top of the cloth. They won't pull the cloth if the object is to one side of it. (In fact, they may give the experimenter a definite "Are you kidding?" look.) This greater understanding of physical causality means their actions look much less magical and are much more effective. This allows them to really plan and scheme and use physical objects as tools.

By the time babies are eighteen months old, they understand quite complicated things about how objects affect each other. Alison and Andy showed babies a toy that was out of reach and then offered them a toy rake. Younger babies would try to reach directly for the object or else would flail around with the rake more or less randomly. But after they reach the age of about eighteen months, babies behaved very differently. They would reach for the out-of-reach object futilely a couple of times, look with a combination of pleading and indignation at their mothers (who were strictly enjoined not to help), stare at the rake, suddenly give a big smile, and then immediately flip the rake over on its side and use it to snag the toy and pull it toward them. You could practically see the lightbulb switching on over their heads. (Of course, these are the same babies who are going through the terrible twos, so their first use of this newfound knowledge about tools is likely to be to pull down all the forbidden objects you put on the high shelves. It does seem perverse of Nature to endow children with new motivations for mischief just when she also endows them with greatly enhanced abilities to get into trouble effectively. It's sort of like letting teenagers get a driver's license.)

There are other reasons to think that, at about a year, babies understand how objects can influence each other. You can show babies a classic case of "billiard ball" causality: a toy car rolling along and bumping into another toy car, which then moves off. Or you can show them almost the same sequence, with just a slight difference—the first car gets close to the second car, and the second car rolls away, but the two cars don't actually touch. Although this is very similar to the first sequence, it violates a basic causal principle. Usually, at least, objects can't act on each other at a distance. Ten-month-olds look longer at the second scene than the first one. This suggests that they recognize just how peculiar it really is. And this, in turn, suggests that they know something about how objects can causally influence each other, quite independent of their own actions.

Children continue to learn about causal relations among objects throughout their toddler years. Before they are three, children are already giving appropriate explanations about what caused what. They say things like "The bench wiggles because these are loose" or "The nail broke because it got bent." By three or four, they can make quite explicit predictions about how simple mechanical systems will work. For instance you can show them a sort of Rube Goldberg apparatus of pipes and tubes through which a ball rolls. Three-year-olds can predict that the ball will have to travel a certain distance before it can bump into another ball and make the machine move.

So, just as babies start out knowing some things about invisible objects but then learn much more, they also start out knowing some things about how events make other events happen but learn much more. Children seem to start out making some assumptions about how they themselves can influence the world, but they gradually have to learn all the many complex ways in which things in the world can influence one another.

Kinds of Things

Try to just tell someone about the objects around you. You'll find that with very few exceptions (such as people and pets) you do this by saying what kinds of things they are, what kinds of categories they belong to. Here on the table are some sweet peas in a glass, four dollar bills, and a cup of coffee. Just by saying that you've already said that these particular things are like a whole bunch of other flowers, or currency, or beverages. You've said that they belong to a particular category.

But there's a paradox here, one that Plato first articulated. All you ever see are individual objects: this particular sweet pea, this individual dollar bill. There is no "sweet-peaness" or "dollarhood" in the world. So how could it ever be informative to say that this individual thing belongs to this nonexistent, mythical category, when the individual thing itself is all we ever actually experience? Plato himself thought the only answer was that there was another universe, a kind of heaven in which the ideal forms of things, the essential sweet-peaness, the ultimate dollarhood, lived. The individual objects in this universe somehow dimly reflected those forms. (Platonic love, for example, was supposed to be the ideal form to which earthly love aspired. Even Plato's followers were skeptical about that one.) That answer won't do for modern cognitive science: categories can't live in heaven any more than objects can be permanent because God is always looking at them.

Another idea might be that something belongs in the same category as another thing because the two things are similar. But the idea of similarity turns out to be impossible to define with any precision. Each individual sweet pea is, after all, different from every other (this one is lavender and this other is more magenta, this stalk has two blooms while the other has three), and each sweet pea is like each individual dollar in some ways (they both are papery and partly green and curl up at the edges and are viewed with greed by those who feel they can never have enough of them).

In fact, the more you think about categories, the more peculiar and complicated they seem. Scientists are always telling us that things we once thought were in one category are really in another. Whales aren't fish. Pandas keep switching back and forth from the bear to the raccoon category (they seem to be back to being bears now, to the relief of all of us gardeners who hate the thought of finding any type of raccoon cute). We're willing to take the scientists' word for it even if we might not be able to say what it is about a panda that makes it a bear (or not). For most of us, it seems that we think there is some deep but vague underlying nature of an object, some essence that makes it belong to a certain category.

How do we learn all this? Like disappearance and causality, categorization seems to be a particularly important problem for babies in the first three years. Even very young infants already can discriminate between different objects and make generalizations about them in some ways. We saw that babies will get bored if they are shown a succession of similar things and perk up if they are shown something different. That, all by itself, means the babies are already categorizing.

In other respects, though, babies don't seem to understand categories in the same way that we do. We noted before that very young babies will track the trajectory of a moving object and that they pay attention to the principle of common fate. Initially, in fact, they seem to use this principle as their main way of identifying objects. We described how babies predict that objects will stay on the same trajectory and move at the same rate of speed. If a toy car moves behind a screen and emerges at the wrong speed or on the wrong path, babies look back toward the screen, as if they think this is a new car and the original must be there somewhere. They assume that an object that traces a particular path of movement is the same object.

However, there is some surprising evidence that young babies are actually not particularly interested if a blue toy car goes in one edge of the screen and a yellow toy duck emerges at the far edge on the same trajectory! A grown-up would assume the duck that came out was brand-new and the other toy was still there behind the screen. But young babies seem content to think the toy somehow magically became a new kind of thing behind the screen. The particular kind of category-crossing magic trick in which the scarf turns into a dove wouldn't be surprising to them. Although young babies can discriminate between yellow and blue, and between the duck shape and car shape, they don't seem to rely on these features to determine which object this is. By the time babies are a year old, however, it is easy to show that across a wide range of situations they are surprised when the car turns into a duck, which suggests they have developed a new view of categorization.

Babies do other things that suggest they have a new view of categories. Alison and Andy gave babies a mixed-up bunch of objects: four different toy horses and four different pencils. Alison would put her hands palm up on the table and watch what the babies did with the objects. Nine- and ten-month-olds picked up the horses and pencils, played with them, and often put them in her hands, but they did so pretty much at random. But twelve-month-olds would sometimes pick all the objects of one group, all the horses or all the pencils, and put them in a hand or in a single pile on the table. By the time they were eighteen months old, babies would quite systematically and tidily sort the objects into two separate groups, carefully placing a horse in one hand and then a pencil in the other. In one experiment a particularly fastidious and precise little girl (there actually are fastidious eighteen-month-olds) noticed that one of the pencils had lost its point. She looked carefully at both hands and then reached for her mother's hand to make a separate spot for this peculiar and defective object.

By the time they are two or three years old, children already seem to have a deeper conception of what it means for an object to belong to a category. They can go beyond the superficial appearance of an object and comprehend something about its essential nature. And they begin to understand that knowing an object's category lets you predict specific new things about the object. For instance, you can tell three-year-olds some new fact about a particular object, you can point to a rhinoceros and say, "This rhinoceros has warm blood." If you then tell them that another animal is a rhinoceros, they will say that it has warm blood, too. But they won't extend their new discovery to a triceratops, which looks like a rhinoceros, if you describe it as a dinosaur.

In a similar study, Alison invented a machine that lit up when you put certain blocks on it but not when you put other apparently identical blocks on it. Then she showed two-year-olds the way the objects influenced the machine. Finally, she picked up one of the blocks that had made the machine go off and said, "This is a blicket. Can you show me the other blicket?" The two-year-olds picked the other block that had made the machine go off, not the blocks that just looked like the "blicket."

These two studies together suggest that even two-year-olds are in some ways like the scientists who reclassify pandas and whales. They look beyond the superficial features of the object to try to determine the deeper laws that govern what the object will do.

By the time children are three or four, we have quite convincing evidence that they look, literally, under the surface of things. Suppose you show three- and four-year-olds natural-looking objects, like plants or rocks. Then you do a kind of cross section, slicing the objects open to show what they look like inside. The children will say that the objects with the same insides are the same kind of thing, even if they look quite different on the surface. Objects with similar outside surfaces but different insides are not the same kind of thing.

These young children, quite surprisingly, even seem to know some things about how animals and plants differ from rocks. They think that living things are more likely to have highly structured insides, while the insides of rocks are more likely to be uniform. They know that baby animals are the same kind of thing as the animals' parents, even though they look very different. They know that tiger cubs, however kittenish they appear, are the same sort of animal as their large and ferocious mothers and quite different from apparently more similar cute and cuddly puppies. They even seem to have a primitive understanding of heredity—they know that a pig who was raised by cows would grow up with a curly tail, like his biological pig parents, and not a straight one, like his adoptive cow parents. These children have barely reached preschool, yet they already seem to have the rudiments of an understanding of biology.

How Do They Do It?

The question, as always, is how do they do it? The answer, as in the last chapter, is that they are born knowing a great deal, they learn more, and we are designed to teach them.

World-Blindness

In the last chapter we talked about how children with autism seem to be blind to other people's minds. They have great difficulty understanding people and often also have difficulty learning how to use language. Another, even rarer genetically determined disorder, Williams syndrome, presents what is in some ways the opposite picture. Children with Williams syndrome are preternaturally sensitive to other people. They are charming and affectionate, even to strangers, and though their language is initially delayed, they develop surprisingly complicated and fluent speech with quite elaborate syntax. But they are terribly bad at comprehending the physical world. They don't even understand hidden objects, use tools, or sort objects into groups until they are three or four years old, although normally developing children work all this out in infancy. As adults, they often can't make their way across a street safely or figure out how to get home. And while they talk about biological and physical phenomena at great length and in some detail, there is a striking superficiality to what they say. A Williams syndrome teenager who can rattle off the names of one hundred different kinds of animals, including pterodactyls and jaguars, nevertheless may fail to understand simple biological processes like growth, inheritance, and death. (Alison's eight-year-old son, who had just painfully worked out the concept of death and was trying to deal with the psychological consequences, heard a visitor talk about Williams syndrome and wistfully remarked that there might be advantages to having only a limited understanding of biology.)

People who study Williams syndrome children often compare their elaborate, fluent speech to cocktail-party talk, the kind of language we use to establish a social connection with other people rather than to achieve a deeper understanding of the world. Whereas children with autism are clueless and frightened in a social setting, children with Williams syndrome are confident but superficial.

We know even less about Williams syndrome than about autism, and there are still many puzzles about just what capacities are spared or damaged in these children. But Williams syndrome suggests there is some genetic basis for our ability to go beyond the surface of things and come to a deeper understanding of the physical world. This ability may be at least partly separate from our ability to speak and to get along in the social world.

The Explanatory Drive

We saw in the last chapter that babies, like scientists, pay attention to counterevidence when they are trying to construct theories of how people work. But there are other similarities between babies and scientists that become particularly vivid when we consider how babies learn about things. In science, and even in ordinary life, we look beyond the surfaces of the world and try to infer its deeper patterns. We look for the underlying, hidden causes of events. We try to figure out the nature of things.

It's not just that we human beings *can* do this; we *need* to do it. We seem to have a kind of explanatory drive, like our drive for food or sex. When we're presented with a puzzle, a mystery, a hint of a pattern, something that doesn't quite make sense, we work until we find a solution. In fact, we intentionally set ourselves such problems, even the quite trivial ones that divert us from the horror of airplane travel, like crossword puzzles, video games, or detective stories. As scientists, we may stay up all night in the grip of a problem, even forgetting to eat, and it seems rather unlikely that our paltry salaries are the sole motivation.

We see this same drive to understand the world in its purest form in children. Human children in the first three years of life are consumed by a desire to explore and experiment with objects. In fact, we take this for granted as a sometimes exhausting fact of parenting. We childproof our houses and say, with a sigh, that the baby is "always getting into things." Clever mothers from time immemorial have discovered that the best way to get a chance to actually cook dinner is to give the baby free rein in the pots-and-pans cupboard.

From the time human babies can move around, they are torn between the safety of a grown-up embrace and the irresistible drive to explore. A toddler in the park seems attached to his mother by an invisible bungee cord: he ventures out to explore and then, in a sudden panic, races back to the safe haven, only to venture forth again some few minutes later. Indeed, we probably never quite escape the bungee cord even as grown-ups; it seems part of the human condition to be perpetually torn between home and away, the desire for comfort and the dread of boredom, the peace of domesticity and the thrill of adventure.

If you think about it from an evolutionary point of view, children's exploratory behavior is rather peculiar. Not only do babies expend enormous energy in exploring the world, their explorations often endanger their very reproductive success (they do have to make it to puberty in one piece, after all). The explanation seems to be that for our species the dangers of exploration are offset by the benefits of learning. The rapid and profound changes in children's understanding of the world seem related to the ways they explore and experiment. Children actively do things to promote their understanding of disappearances, causes, and categories.

Fortunately, these aspects of the physical world are so ubiquitous that babies can do their experiments quite easily and for the most part safely. The crib, the house, the backyard are excellent laboratories. For instance, we can see babies become interested in, almost obsessed with, hiding-and-finding games when they are about a year old. There is the timeless appeal of peekaboo, that irresistibly funny surefire daddy routine that never seems to go stale. Babies also spontaneously undertake solo investigations of the mysterious Case of the Disappearing Object. Alison once recorded a baby putting the same ring under a cloth and finding it seventeen times in succession, saying "All gone" each time. In our experiments, babies often begin by protesting when we take the toy to hide it. But after one or two turns, they often start hiding the toy themselves or give the cloth and toy to us with instructions to hide it again. Eighteen-month-olds, who are not renowned for their long attention span, will play this game for half an hour.

Babies are similarly fascinated by causal relations between objects. Babies in the ribbon-and-mobile experiments actually get bored after a while with the spectacle of the mobile moving, but they don't get bored with the sensation of their own power. After a while, they only occasionally glance at the mobile, but they keep on kicking. The ubiquitous baby "busy box" is another toy that depends on babies' fascination with what happens in the world. By the time babies are one or two years old, they will quite systematically explore the way one object can influence another object. The babies in our rake experiments forget all about getting the toy after a trial or two. They often deliberately put the toy back far out of reach and experiment with using the rake to draw it toward them. The toy itself isn't nearly as interesting as the fact that the rake moves it closer.

Similarly, babies persistently explore the properties of objects. Six- or seven-month-olds will systematically examine a new object with every sense they have at their command (including taste, of course). By a year or so, they will systematically vary the actions they perform on an object: they might tap a new toy car gently against the floor, listening to the sound it makes, then try banging it loudly, and then try banging it against the soft sofa. By eighteen months, if you show them an object with some unexpected property, like a can that makes a mooing noise, they will systematically test to see if it will do other unexpected things. And, as we saw, children this age will quite spontaneously sort different kinds of objects into different piles.

We think this kind of playing around with the world actually contributes to babies' ability to solve the big, deep problems of disappearance, causality, and categorization. Before science became a separate, socially defined field, it was called experimental philosophy. The grown-ups who chat about real-estate prices while the baby is, thank heaven, busy playing with her toys, don't realize they are witnessing positive miracles of experimental philosophy.

Grown-ups as Teachers

The grown-ups, though, like biblical shepherds, may be contributing to the miracle even if they don't quite recognize it. When babies are about a year old, grown-ups start talking to them in a distinctive way. They start giving a sort of sportscaster's play-by-play of everything the baby does. "There, you picked up the cup, oh, now you're putting it down again. Whoops, there it goes. Oh, dear." And so on. It may not seem quite as silly as the "Aren't you a pretty bunny" talk, but if you think about it, it still seems pretty silly. It's not, after all, as if you're telling the baby something he doesn't already know.

The silliness, however, may be only on the surface. We have reason to think this kind of early language helps organize the world for babies. In the last chapter, we described a sort of natural experiment to test the effects of other people on children's understanding of the mind: we compared children with older siblings to those without them. We can do a similar sort of natural experiment by comparing children who hear parents describe the world in different ways.

It turns out that, just by the nature of the grammar of their languages, Korean- and English-speaking parents talk about the world quite differently. Korean (like Latin or French) uses an elaborate system of different verb endings to convey different meanings. As a consequence, Korean-speaking parents can, and often do, omit nouns altogether when they talk to their children. A Korean mother can say the equivalent of "moving in" when she sees the baby put a block in a cup, without saying anything about who or what is doing the moving or what it's moving into. In English, on the other hand, we must include at least one noun in almost every intelligible sentence. Moreover, English-speaking parents spend a lot of time pointing to objects and giving them names: "There's a dog! Look at the bird! Car! Airplane!"

Alison and a Korean colleague, Soonja Choi, looked at the kinds of things English-speaking mothers and Korean-speaking mothers said to their eighteen-month-old babies and found that this was indeed true: English-speaking mothers used more nouns and fewer verbs than Korean-speaking mothers. English-speaking mothers tended to name objects a lot, while Korean-speaking mothers were more likely to talk about actions.

When Alison and Soonja looked at what the eighteen-month-old children understood about the world, they found there were consistent differences between the Korean and English speakers. Like their parents, the Korean children used more verbs than the English-speaking kids, while the English-speaking kids used more nouns. But in addition, the Korean-speaking children learned how to solve problems like using the rake to get the out-of-reach toy well before the English-speaking children. English speakers, though, started categorizing objects earlier than the Korean speakers. For instance, they were more likely to put the toy horses and the pencils into two separate piles. It was as if the Korean-speaking children paid more attention to how their actions influenced the world, while the English-speaking children paid more attention to how objects fit into different categories. The likeliest explanation for this is that the children were influenced by what the grown-ups around them said, which in turn was shaped by the grown-ups' language.

This may sound a bit more radical than it actually is. Many years ago the linguist Benjamin Lee Whorf suggested that the grammar of our language influenced the way we thought. The Whorfian hypothesis, as it is called, fell into scientific disrepute quite quickly. (It kept a lasting appeal in the popular imagination, however. In the 1980s a high American official said the Russians would never really negotiate a peace because they didn't even have a word for détente in their language. The fact that *détente* is a French word didn't seem to occur to him.) For one thing the idea is logically incoherent. How do we know that another language even has a concept that ours doesn't have, unless we can somehow express that concept in our language, too?

What we found (and a number of other people have recently found) is rather different from Whorf's idea. Both the Korean- and the English-speaking children understood actions and object categories by the time they were two. Still, the difference in emphasis in the two languages seems to have made one problem easier to solve for one group, while the other problem was easier for the other group. It's like the difference between children who grow up in a house where they talk about music all the time (like Andy and Pat's house) and children who grow up in a house where they talk about politics (Alison's husband is a public-radio journalist). The children in each home may have the capacity to understand music or politics, but they naturally know more about the topic they hear about a lot.

The interesting thing is that even quite tiny children, just beginning to say their first words, already seem to be influenced by what the people around them are talking about. And, of course, the parents are exercising this influence completely unconsciously, just by talking to their babies. In fact, it would probably be impossible for an English-speaking parent to consciously start to talk like a Korean speaker or vice versa. Nature wisely doesn't rely just on the conscious resolutions and good intentions of parents. The most potent influences on babies—the nouns they hear in a sentence or the unwitting lessons of their siblings—are influences that no one consciously wields.

Babies seem to learn about the external world in much the way they learn about other minds. They start out with some crucial assumptions, assumptions that seem to be built in. But, just as important, they are endowed with powerful abilities to learn, and even more powerful motivations. They are as driven to explore the alien physical world around them as they are to make first contact with the local species. A one-year-old set loose to crawl around a new living room will have the unmistakable gleam in her eye of one who boldly goes where no one has gone before. Fortunately, the life-forms they encounter are benign and quite genuinely, if unconsciously, dedicated to bringing the fruits of their wisdom and civilization to these intrepid small voyagers.

CHAPTER FOUR

What Children Learn About Language

The problem of Language, like the External World problem, is largely invisible in everyday life. While we know abstractly that we are hearing a sequence of arbitrary sounds, it feels as if thoughts are simply pouring into our minds. Suppose your spouse were to come into the room now and say, "You know, that woodwork could use some sanding." It would take, at most, a second or two for you to understand that sentence. Yet during that second you would have to do a remarkably complex set of calculations.

First, you have to break up the continuous stream of sounds into separate pieces and identify each sound accurately. Very small differences in sounds can make a big difference in meaning; *you know* means something quite different from *you go* or *who knows.* Then you have to string the sounds together into words. Because the average English speaker knows more than seventy-five thousand words, there are a lot of possibilities. Then, assuming you know the words, you have to combine them to make a sentence. Just as small differences in sound can make a big difference in meaning, so small differences in the arrangement of words can make a big difference in meaning. The sentence "You know, could this sand use some woodworking?" would mean something quite different from your partner's remark. (It is one of life's tragedies that "John loves Mary" does not mean the same thing as "Mary loves John.") Then you need to understand all the nuances of meaning each word can have. You need to know that the word *sand* in that sentence refers to an action and not the stuff on the beach, and that *you know* doesn't actually refer to your knowledge at all. And, finally, you have to figure out something about the larger intent of the sentence. Is your partner reproaching you for reading instead of doing household chores? Or announcing an intention to spend the next hour sanding the woodwork, so you might want to move to a less noisy location to read? You figure all this out instantly and without any conscious effort.

Just as a magic show makes us realize how much we take for granted about things, visiting a foreign country makes us realize how much we take for granted about words. You greet your spouse's comment with effortless understanding, but if he were speaking a foreign language, you would feel baffled incomprehension instead. One of the brilliant aspects of the film *The Third Man* is that the actors who play the inhabitants of postwar Vienna actually speak German, with no translations or subtitles. When you watch the film, you find yourself experiencing just the same vertiginous incomprehension as Joseph Cotten, the innocent American hero. Simply hearing what are clearly important sentences spoken in a strange language rocks our usual calm assurance that we have some idea about what's going on (even without Orson Welles looming out from behind bullet-riddled baroque statues in the background). It isn't just that you don't know what the words mean; you don't even know what the words, or even the sounds, are, or where one sound ends and the next begins. Everyone seems to talk so fast. This, of course, is where babies start out. In fact, the babies are worse off than Joseph Cotten in some ways, because they have no other language in which they can express their bafflement.

The Sound Code

Learning to understand a language is like cracking a deeply encrypted code. We all crack this code effortlessly, at an age we can't even remember, and we use it effortlessly as adults. But it turns out that the code is far more baffling than any spymaster's cryptogram. No computer has been able to figure it out.

When people comment on the scientific impossibilities of *Star Trek*—light speed and warp drive and even holodecks and replicators—they rarely mention what seems like a small technological detail. On *Star Trek* people talk to the ship's computers and the computers understand them (in fact, they even talk to the ship's doors). That technology may not be quite as distant as warp drive, but it's not close. Today, computer software companies all over the world are trying to create a machine that can understand spoken language. Companies and governments have spent billions of dollars on speech technology over the last fifty years, but no computer in the world has solved the Language problem yet. Our bathroom scales and elevators can produce a bit of understandable if annoyingly unnatural speech, but there is no computer that can do what every three-year-old can do: understand a conversation.

To nonscientists—even to the guys investing the billions of dollars—it isn't obvious why the problem is so hard. Pat was returning home from a conference in 1991, and, conquered by the usual airplane exhaustion, she slumped into her seat next to a young guy with a backpack. The young guy turned out to be her fellow Seattleite Bill Gates, CEO of Microsoft. Pat spent the next four hours answering Bill's questions about why it was so difficult for his computers to understand speech. At the time, Microsoft and other computer companies were struggling to liberate computer users from keyboards. What the science could tell them was largely why the problem was so hard, rather than how to solve it.

The core of the problem, as in the problem of the External World and that of Other Minds, is the mysterious gap between the sound waves that actually reach our ears and the sounds and words we create in our minds. We can make a sort of photograph of a sound called a spectrogram. The spectrogram shows the actual physical properties of the sound waves: how loud they are, what pitch they are, and how they change. Just as we must translate the two-dimensional pattern of light on our retinas into the three-dimensional solid objects we perceive, so we must translate this pattern of sounds into language. The distance from there to here is just as great.

There are some glaring problems that become obvious as soon as you compare the spectrogram with the words we perceive. First, the sounds of human speech aren't like beads stacked next to one another on a string: there are no gaps or pauses between the sounds in the spectrogram. Instead, the sounds are continuous, and we have to divide them into units. Second, all voices are different because our mouths are all different sizes and shapes, so even simple sounds (like *ah*) sound different depending on who says them. And when we speak more quickly or more slowly, which we do all the time, the sound waves change again. Moreover, each time a consonant sound, such as *b* or *d*, is placed in front of a different vowel, the sound changes. The *d*'s in front of the words *dude* and *deed* are so different that a spectrogram of the *d* in *dude* actually looks more like a *g* than like the *d* in *deed*.

Finally, and most complicated of all, people speaking different languages hear sounds totally differently. A sound with exactly the same spectrogram will be heard differently by someone speaking Japanese and someone speaking English. Two physically different sounds (like *r* and *l*) may sound identical to a Japanese speaker but completely different to an English speaker. It isn't only that you must figure out how to get from the sensations to the representations, as you do when you translate the two-dimensional retinal images to a three-dimensional world. You also must do that translation differently for each different language.

Three-year-olds have solved all these problems. They can recognize a *d* sound whether it's spoken by Mom or Dad, whether it's in *deed* or *dude*, whether it's quickly whispered or slowly sung, and they make just the right discriminations for English. Computer systems that can do some speech recognition can't match the three-year-old. As we mentioned, most English speakers know more than seventy-five thousand words. If you limit your speech to the ten digits of telephone numbers, or even to the cities in America, computer speech recognition functions well. But real conversations can't happen using ten words, or even a thousand words.

One of the biggest problems for computers is segmenting speech into separate units for analysis. Early computer programs solved this by having speakers separate each word. People using this voice-recognition technology had to speak very slowly, SEPARATING (one-second pause) EACH (one-second pause) WORD (one-second pause) WITH (one-second pause) A (one-second pause) VERY (one-second pause) ANNOYING (one-second pause) ONE- (one-second pause) SECOND (one-second pause) PAUSE. To solve the problem of different voices, computers are programmed to recognize only one particular person's voice and then have to be reprogrammed for each separate user. In the same way, computers are programmed to treat the *d*'s in *deed* and *dude* separately, as if each were a completely different sound. In 1998 the first continuous speech recognition software for dictation became available, but it still requires training for each separate user and limits your vocabulary.

So much for the notion of a *Star Trek* computer that answers your every question accurately in a soothing if somewhat chilly voice. By the time any current computer could understand "Implement evasive maneuver Alpha Theta and fire photon torpedoes at Romulan vessel on my mark," the *Enterprise* would be a noncorporeal energy particle pattern.

Making Meanings

All this complex code breaking is necessary just to figure out the words of a language from the sounds you hear. This part of understanding language happens so effortlessly and so quickly that it's hard even to recognize the problem at first, and scientists have only started tackling it quite recently. The next part of the process, getting from words to meanings, is more obviously difficult. Philosophers have pondered for millennia the question of how words can mean things.

It almost seems as if there is a magic link between the words we use and the outside world. Say a word and suddenly you are in touch with the thing the word refers to, no matter how distant or strange. Many cultures and religions explicitly believe in word magic; knowing the true name of a thing gives you power over it. But when you think about it, our everyday language is just as mysteriously powerful. Consider what is happening as you read this book. By casting your eye over a bunch of arbitrary printed shapes, you are suddenly in contact with the minds of three people thousands of miles away from you. You visit a laboratory you've never seen and meet children who grew up long ago. And as every novel reader or letter writer or Internet cruiser knows, words can not only take you to other worlds, they can create worlds of their own. You don't need *abracadabra; once upon a time* will do. How can words defy all the limits of space and time and possibility in this way? How could anyone learn to wield this kind of power?

Almost two millennia ago Saint Augustine proposed one solution to the problem, perhaps the most obvious one: as children, we heard our parents say words and point to things, and we associated the words with the things. But the more you think about it, the less adequate this solution seems. Over the centuries other philosophers have demonstrated the difficulties. Bertrand Russell showed that meanings go beyond just the things we point to. How do we learn words like *unicorn,* words that refer to things that don't exist? How do we learn all the words—verbs and adjectives and prepositions—that don't refer to things at all? Ludwig Wittgenstein raised another set of questions. How do we learn not just what the words refer to but what the speaker wants us to do about them? After all, even to understand pointing you need to know something about the intention of the person who points. You have to know that the gesture of extending your index finger is a way of picking out an object to be named rather than, say, casting a curse or conferring a blessing. The philosopher Willard Quine raised yet another set of questions. How do we know that a name refers to the thing someone points at rather than to the thing plus a bit of the empty space near it or some part of the thing? Computers still can't solve the problem of decoding sounds, but they are even further away from solving these sorts of problems.

The Grammar We Don't Learn in School

In addition to the challenge of decoding sounds and meanings, there is still another set of problems. In the 1960s Noam Chomsky raised a whole new set of questions which people hadn't paid much attention to earlier. How do we combine words to make new sentences? Almost all of the sentences we hear are brand-new, and yet we have no difficulty figuring them out. Even when we know individual words, arranging those words in different ways can lead to different meanings. Chomsky's answer to that question created a new field, modern linguistics, and a new way of thinking of the old idea of grammar.

Traditional grammar, the sort of thing we used to learn in school, vacillated between telling you what speakers of a language actually did and saying what speakers ought to do. Chomsky argued that knowing a language involved knowing a set of unconscious rules, but they weren't like the rules of traditional grammar. These rules weren't socially imposed like the rules of traffic codes or Monopoly or the old elementary school grammar books. Rather, they were natural, unconscious rules. They were like the rules we use when we turn visual information into representations of objects. They were more like the law of gravity than like the law of the land (or the law of the English teacher).

Chomsky's solution to the problem of Language is much like the modern solution to the External World and Other Minds problems. We are designed to take in sequences of sounds and translate them into representations of meanings just as we are designed to take in sensory information and translate it into representations of objects and to take in facial expressions and translate them into representations of feelings. We have an implicit set of rules that allows us to transform the sequences of sounds we hear into sequences of ideas.

We actually know quite a lot about how some parts of this system work. For instance, we know a great deal about how we translate the sounds we hear into meaningful units like words, though still not enough to let us program a computer to do it. We know something, though less than Chomsky originally hoped, about how we combine words to make sentences. Other parts of the problem remain deeply mysterious. In particular, the problem of how words come to refer to the external world, of how meanings are made, is almost as mystifying to contemporary linguists as it was to Saint Augustine.

We assume that this system was designed by evolution, and it is certainly distinctively human. Perhaps the most obvious advantage of language is that it lets us communicate and coordinate our actions with other people in our group. But language also has other less obvious but equally distinctive advantages. The fact that we speak different languages also lets us differentiate between ourselves and others; it's as good a way as any of knowing who is part of your group and who is an outsider (keeping information away from your enemies may be almost as important as sharing it with your friends). And the development of language is probably linked to the development of our equally distinctive ability to learn about people and things. It allows us to take advantage of all the things that people before us have discovered about the world. We can see so much further than other species because we stand on the shoulders of our mothers and fathers (who at least look like giants to babies).

Chomsky's solution raises a deeper developmental problem: Where does this linguistic system come from? By the time they are in kindergarten, children have mastered almost all of the complexities of their particular language, with no conscious effort or instruction. How do they do it? The broad lines of the developmental answer to this question should be familiar by now. Babies are born knowing a great deal about language. They also have powerful learning procedures that allow them to add to that knowledge and, in particular, to learn all the details and peculiarities of the language of their own community. Finally, adults play an especially crucial role in language learning.

The analogy to science works very well when it comes to explaining how babies solve the Other Minds problem and the External World problem. There really is a world of objects and minds out there. Babies make up theories about that world, but those theories can always be revised if new evidence comes along. In the case of language, however, the problem is rather different. It is not about discovering an independent reality but about coordinating what you do with what other people do. There isn't any abstract "language" out there that is independent of what people say. We could find out (in fact, we do find out) that we are all wrong about some important aspect of the world or other people. But we couldn't find out that we are all speaking English the wrong way; English just is the language we speak. So the babies' Language problem is not so much the scientist's problem—find out what the world is really like—as it is a kind of sociological or even anthropological problem: find out what the folks around here do and learn how to do it yourself. The other folks are crucial.

The problem is difficult because different communities speak different languages, sometimes quite radically different languages. Babies don't know beforehand which language they are going to be exposed to. Potentially, they have to be able to master any one of thousands of different languages. And yet, by the time they are four or five, children have figured out precisely which language is spoken in their community.

Grown-ups are both the teachers and the subject matter.

What they say is the only source of evidence about what the language is like. And for the children the aim of the enterprise is not just to find out about the grown-ups' language but also to make that language their own.

What Newborns Know

Ask anyone when children start to learn language. Almost everyone will say that language begins when babies say their first words. But the new techniques for uncovering what babies know have led to a surprising discovery. Babies know important things about language literally from the time they are born, and they learn a great deal about language before they ever say a word. Most of what they learn at that early age involves the sound system of language. We decode the sound cryptogram, and solve many of the problems that still baffle computers, before we can actually talk at all.

We mentioned that part of what makes learning language difficult is that languages carve up sounds and different languages carve them up differently. A wide variety of different sounds, with very different spectrograms, will all seem like the same sound to us, and, in turn, that sound will seem sharply different from other sounds that are actually quite similar to it physically. Suppose you use a speech synthesizer to gradually and continuously change one particular feature of a sound, such as the consonant sound *r,* and play that gradually changing sound for people. You very gradually and continuously change the *r* sound to *l.* What is actually coming into the listeners' ears is a sequence of sounds, each of which is just slightly different from the last. But what they perceive is someone saying the same sound, *r,* over and over, and then suddenly switching to a new sound, *l,* over and over. The listeners have divided up the continuous signal into two sharply defined categories: either it's an *r* or an *l,* not anything in between. They can't distinguish between all the different *r*'s, even though the sounds themselves are quite different. Scientists call this categorical perception, because a continuously changing set of sounds is perceived categorically as being either black or white, *r* or *l,* with nothing in between.

The way we categorically perceive speech is unique to each language. In English we make a sharp categorical distinction between *r* and *l* sounds. Japanese speakers don't. In fact, Japanese speakers can't hear the distinction between American *r* and *l,* even when they are listening very hard. (Hence all the dubious jokes about Japanese speakers ordering what sounds like "flied lice" instead of "fried rice.") Pat was in Japan to test Japanese adults and their babies on the *r-l* distinction. She had carefully carried the computer disk with the *r* and *l* sounds to Japan, and when she arrived in the laboratory in Tokyo, she played them on an expensive Yamaha loudspeaker. She thought that such clearly produced sounds would surely be distinguished by her Japanese colleagues, who were quite good English speakers as well as being professional speech scientists. As the words *rake, rake, rake* began to play out of the loudspeaker, Pat was relieved to know that the disk worked and the sound was perfect. Then the train of words changed to an equally clear *lake, lake, lake,* and Pat and her American assistant smiled, looking expectantly at her Japanese colleagues. They were still anxiously straining to hear when the sound would change. The shift from *rake* to *lake* had completely passed them by. Pat tried it over and over again, to no avail.

This happens to all of us, of course, as we try to hear distinctions that are used in languages that are not our own. Say we use the speech synthesizer again to change the sound *b* gradually and continuously to *p,* and we test speakers of many languages on these sounds. Americans will hear two sharp categories, *b* and *p*. Spanish and French speakers listening to these same sounds hear two categories also but divide the continuous stream of sounds in a different place than the Americans do. What sounds like *b* to a Spanish speaker will sound like *p* to an English speaker. Thai speakers hear three categories. In each case listeners hear sharp changes—quantum leaps—between categories, with no in-between. But the speakers of Spanish, French, and Thai hear those quantum leaps at different places. And we English speakers don't even notice the categorical shifts Spanish or French or Thai speakers hear, just as the Japanese speakers didn't notice the change from *rake* to *lake*. It's as if the speakers of each language have a very different way of transforming the actual sound waves that come into their ears into a set of language sounds.

Why do the speakers of different languages hear and produce sounds so differently? Ears and mouths are the same the world over. What differs is our brains. Exposure to a particular language has altered our brains and shaped our minds, so that we perceive sounds differently. This in turn leads speakers of different languages to produce sounds differently. When and how do babies start to do this? Do they start out listening like a computer, with no categorical distinctions? Or do they start out with the categorical distinctions of one particular language, say English or Japanese or Russian?

We can't ask babies directly whether they think two sounds are the same or different, but we can still find out. Very young babies can tell us what they hear by sucking on a special nipple connected to a computer. Instead of producing milk, sucking on this special device produces sounds from a loudspeaker, one sound for each hard suck. Babies love the sounds almost as much as they love milk: they may suck up to eighty times a minute to keep the sounds turned on. Eventually, though, they slow down; they get bored hearing the same thing over and over again. When the sound is changed, however, infants immediately perk up and suck very fast again to hear the new sound. That change in their sucking shows that they can hear a difference between the new sound and the sound they heard before. Using this technique we can do the same *r* and *l* experiment we just described with adults. We can use a speech synthesizer to present the babies with a slowly and continuously changing consonant sound. Then we can test the babies to see which sounds they think are the same and which sounds they think are different.

Scientists anticipated that these tests would show that very young babies initially can't hear the subtle differences between speech sounds and only slowly learn to distinguish those that are important in their particular language, such as *r* and *l* in English. In fact, the results were just the reverse. In the very first tests of American infants listening to English, babies one month old discriminated every English sound contrast we threw at them. Moreover, the babies demonstrated the categorical perception phenomenon. They thought all the *r*'s were the same and different from all the *l*'s, just as adult English speakers do.

But then shortly afterward speech scientists discovered something even more remarkable. Kikuyu babies in Africa and Spanish babies in Mexico were also excellent at discriminating American English sounds as well as the sounds of Kikuyu and Spanish, and American babies were just as good at discriminating Spanish sounds—much better than American adults. The sophisticated Japanese scientists who strained to hear the difference between *rake* and *lake* would not have had any trouble doing so when they were forty or fifty years younger. Very young babies discriminated the sounds not only of their own language but of every language, including languages they had never heard. Infants were as good at listening to American English as they were at listening to African Kikuyu, Russian, French, or Chinese regardless of the country they were raised in. Pat also discovered that babies, unlike computers, make these distinctions no matter who is talking—a man or a woman, a person with a high squeaky voice or one with a deep resonant voice.

So babies start out knowing much more about language than we would ever have thought. Newborn babies already go well beyond the actual physical sounds they hear, dividing them into more abstract categories. And they can make all the distinctions that are used in all the world's languages. Babies are "citizens of the world." Perhaps we grown-up scientists failed to predict this because our skills are so much more limited. Our citizen-of-the-world babies clearly outperform their culture-bound parents.

Taking Care of the Sounds: Becoming a Language-Specific Listener

Providing the answer to one puzzle creates another. If babies are born listening like universal linguists, how do they grow up to be culture-bound language specialists? Japanese babies learn English if they're raised in America and Japanese if they're raised in Japan. When does a Japanese baby learn that in her language it doesn't matter whether she produces *r* or *l* because the adults in her culture can't hear the difference anyway and it won't change the meaning of the word in Japanese? When does the American baby learn that the difference between the two Spanish *b*'s doesn't matter in English? Most scientists thought at first that babies would appreciate these language-specific distinctions only after they had already learned quite a lot of meaningful language.

To answer this question we needed a way to test babies once

they had listened to their particular language for a while. After four months or so, many babies are much less likely to suck to turn on computer sounds, so that technique doesn't work as well. But another test works very well with six- to twelve-month-olds. The babies sit on a parent's lap. On their right a person keeps their interest by slowly manipulating toys—dangling a plastic spider, turning a toy horse upside down, and doing other visually interesting things. On the babies' left there is a loudspeaker with a black box on top of it. A sound repeats out of the loudspeaker, something like *oo, oo, oo.* Every once in a while the sound is changed to *ee, ee, ee.* When babies hear the sound change, they tend to get distracted from the interesting person and look over toward the loudspeaker. When they do, the black box on top of the loudspeaker lights up. Inside the box a bear dances or a monkey pounds a drum, delighting the babies. Then the display goes off and the babies turn back to the interesting person on their right. Soon the babies figure out that if they turn their heads toward the loudspeaker when the sound changes, they'll see something interesting. Whether they turn their heads or not tells us whether they heard the sound change or not.

When Pat went to Japan to test adults on the American *r* and *l* sounds, she also tested babies. Japanese and American seven-month-olds discriminated *r* from *l* equally well. But just three months later, the two groups of infants were as different as night and day. At ten months, Japanese infants could no longer hear the change from *r* to *l.* American infants not only could do so but had actually gotten much better at making this distinction. A previous study of babies being raised in English-speaking homes had similar results. That study showed that at six months Canadian babies could discriminate Hindi speech sounds that Canadian adults can't distinguish. But by twelve months the Canadian babies could no longer do so.

As they hear us talk, babies are busily grouping the sounds they hear into the right categories, the categories their particular language uses. By one year of age, babies' speech categories begin to resemble those of the adults in their culture. Pat conducted some even more complicated experiments with Swedish babies using simple vowels to see how early they start organizing the sounds of their language in an adultlike way. She showed that at six months the process has already begun. The six- to twelve-month time span appears to be the critical time for sound organization.

What might be happening to the babies between six and twelve months? One way of thinking about it is in terms of what Pat calls prototypical sounds. After listening to many *r* sounds in English, for example, babies develop an abstract representation of *r*—a prototypical *r*—that is stored in memory. When we want to identify a new sound, we seem to do it by unconsciously comparing the new sound to all of the prototypes stored for our language and picking the one that's the best overall fit. Once we've unconsciously done this, we distort the way we hear a sound to make it more like the prototype stored in memory than like the sound that actually hits our ears.

It's similar to what happens when you show people a drawing of something they've seen very often, a house, for example, and then ask them to copy it from memory. If the house you show them doesn't have a chimney, many people will add one to their drawing anyway, even though it wasn't in the original drawing they saw. Once they coded the picture as a house, they distorted their memory of it to make it more like what they think of as the prototypical house. We can do complicated analyses to show just what the prototypes of our speech sounds are and just how we distort what we hear to suit them. Our language prototypes "filter" sound uniquely for our language, making us unable to hear some of the distinctions of other languages. Pat's tests suggest that babies' language prototypes begin to be formed between six and twelve months of age.

It isn't just that younger babies have a skill they lose later on. Rather, the whole structure of the way babies organize sounds changes in the first few months of life. Before they are a year old, babies have begun to organize the chaotic world of sound into a complicated but coherent structure that is unique to their particular language. We used to think that babies learned words first and that words helped them sort out which sounds were critical to their language. But this research turned the argument around. Babies master the sounds of their language first, and that makes the words easier to learn.

When babies are around a year old, they move from sounds to words. Words are embedded in the constant stream of sounds we hear, and it is actually difficult to find them. One problem computers haven't yet solved is how to identify the items that are words without knowing ahead of time what they are. Try to find the words in a string of letters like *theredonateakettleoftenchips.* The string contains many different words: *The red on a teakettle often chips* or *There, Don ate a kettle of ten chips* and so on. Of course, in written language there are normally spaces between words. But in spoken language there aren't actually any pauses between words. That's why foreign speech sounds so fast and continuous, and that makes the Language problem very hard for computers to solve.

Babies seem to learn some general rules about the words in their particular language before they learn the words themselves. By nine months, for example, they've learned that English contains words that have a certain emphasis pattern: words with a first-syllable stress pattern, like *BASEball* or

POPcorn, are more common than the reverse (a word like *surPRISE*). In some other languages it's the other way around: first syllables are stressed less often than last syllables. By nine months babies have all this sorted out. American babies prefer to listen to words with the American pattern, while babies from other countries prefer to listen to words typical of their own languages, even though the babies at this age don't understand the meanings of these words.

After babies learn which sounds are possible in their language, they learn which sound combinations are possible. In English, for instance, the sound combination *zb* is not possible. No English word contains this combination. In Polish, however, this combination is possible. (Zbigniew Brzezinski, President Carter's national security adviser, could never have become president because, for one thing, hardly any Americans could pronounce his name.) By nine months babies show a preference for listening to sound combinations that are possible in their language, even if the sound combinations don't form real words in the language but are only potential words. American nine-month-olds already have trouble with words like *Zbigniew,* while Polish nine-month-olds would think it's no problem. Knowing which words are possible in your language helps you begin to divide the continuous stream of speech into words, even if you don't know what those words mean. If you are American, you can already eliminate the strings that have *zb* in them or that have the wrong stress pattern.

The Tower of Babble

So babies are learning about speech a long time before they begin to talk. But, of course, what parents notice most is what babies actually say. Whether babies are born in Paris, Zimbabwe, Berlin, or Moscow, they start to coo when they're about three months old. They make delightful little *oohh* and *aahh* sounds when a parent is face-to-face with them, talking and smiling. Babies seem to grasp intuitively that humans take turns in this kind of exchange. They coo, we goo, and thus we have our first conversations with our children. Babies already know something about how dialogue works.

A short time later, at about seven or eight months, babies begin to babble. They start producing strings of consonant-vowel syllables, *dadadada* or *babababa.* Babies across cultures babble at first in an identical way, producing consonant-vowel combinations using sounds like *b, d, m,* and *g* with the vowel *ah.* Pat and Andy vividly remember when their daughter, Katherine, started to babble. As a speech scientist Pat had waited for months for the moment babbling began. She'd even set up a recorder hoping to catch it on tape. One morning Pat took Kate out for a walk in the stroller and stopped at the local Starbucks for a latte (it's Seattle, after all). Kate was happily cooing and gurgling at the line of customers as Pat ordered, when suddenly a "Babababa" rang out. Pat froze and waited to hear it again, asking the cashier and other Starbucks habitués to listen for confirmation. With the customers poised on the edge of their seats, Kate blissfully went on, "Babababa." She babbled right on time, like clockwork, like all babies around the world.

Once babies reach the babbling milestone, the universal phase of language production ends. Babies from different cultures, learning different languages, start to make the distinctive noises of their own community sometime between a year and a year and a half. The Chinese baby starts to babble in a way that sounds Chinese. She uses very rapid pitch changes just like adult Chinese speakers. Swedish babies babble in a way that sounds distinctly Swedish, using the rising intonation patterns typical of adult speakers of Swedish. (They sound a bit like the Swedish chef on *The Muppet Show.*)

The First Words

So far we've talked about how children come to understand the system of sounds of the language they hear. This apparently simple problem turns out to be extremely complicated. Babies are hard at work on it throughout infancy. But we haven't even started in on what most of us think of as the central problem of language: learning what words mean.

Remember that Augustine thought there was a simple answer to this problem: children saw things and heard their parents name the things and then associated the things with the names. That idea has a strong intuitive appeal. If you ask the average parent, or for that matter the average psychologist, when babies begin to talk, they will tell you about when they use their first names. Usually, parental egocentricity being what it is, the parents report "Mama" and "Dada." It turns out, in fact, that across a variety of very different languages the "baby words" for mother and father are very similar; "Mama" and "Dada" are joined by "Mati" and "Tata." They are also, of course, precisely the sounds that babies are very likely to produce spontaneously when they babble. So it's not entirely clear whether babies say "Mama" and "Dada" because that's what their beloved parents call themselves, or whether parents call themselves Mama and Dada because that's what the babies say anyway.

While philosophers, psychologists, and parents were so sure they knew how babies started to speak, no one consulted the babies themselves until around the 1970s. With the advent of videotape you could actually watch what children said and when they said it. The results were surprising. Babies did say "Mama" and "Dada" (parents weren't utterly self-deluded), and also "juice" and "ball" and "doggie." But they said many other things that grown-ups didn't notice. Perhaps parents didn't notice because the words children used were so peculiar. Babies consistently said things like "gone," "there," "uh-oh," "more," and "what's that?" among their very first words. Why these rather odd words? And what did they mean to the babies?

Alison set out to find out. She had come to Oxford to study philosophy. But she was interested in how language begins, so she spent many hours each week watching babies in the big, drafty rooms of North Oxford villas. The grand houses have been divided into apartments, but they still have some of the same atmosphere they had when Lewis Carroll told stories about Wonderland to Dean Liddell's daughters. Oxford is still gray and wet and dim and full of elegant, chilly buildings and faces. There is still nowhere more elegant and more chilly than Logic Lane, where the philosophy classes are held. And yet Oxford also still has lamplit rooms full of luminously red-headed toddlers gathered around tables of cream cakes and milky tea, and little courtyard gardens with iron gates where pink-cheeked three-year-olds play ball. The children had much the same effect on Alison, a young American woman philosopher, that they had had on Lewis Carroll, the elderly bachelor logician, a hundred years earlier. They were a bright glimpse of clarity and warmth compared with the vaguely threatening and deeply eccentric creatures of Logic Lane. And yet, at the same time, the private children's world had its own mystery and strangeness, even if the strangeness often seemed more sensible than the accepted craziness of the world outside.

For while the babies' language was superficially simpler than the convoluted paragraphs of Logic Lane, it was, in its own way, just as peculiar. Almost always your first guess at what a word meant would turn out to be wrong. For example, *gone* was one of the most common words these babies used. When parents notice this at all, they assume that it has something to do with finishing up food. But, in fact, the babies rarely used the word this way. In one taped hour Henry, a particularly cherubic eighteen-month-old, said "gone" turning over a small piece of wax paper with a bit of brown sugar attached to it so that the sugar became invisible (seven times in a row), turning the page of a picture book so that each baby animal was no longer in sight, hiding a ring under the edge of Alison's skirt (twelve times in a row), putting a block inside a toy mailbox (seven times in a row), and plaintively searching for Mum, who had gone to a neighbor's ("Mummy gone!"—think Jackie Coogan in *The Kid*).

It turned out that *gone* didn't have much to do with food at all, or with the way grown-ups use *gone.* Instead, Henry, and the other babies Alison studied, used *gone* to describe the many and varied ways that objects disappear from view. They commented on the fact that they couldn't see something they knew was still out there somewhere.

Alison discovered that there were a number of other words that seemed to work the same way as *gone.* For instance, babies often use a word to indicate whether they succeed or fail in doing something. American babies use *there!* to note their successes and *uh-oh* to describe failures, while the babies in the Oxford villas used the more genteel *oh dear* (although one British baby did briefly but memorably say *oh bugger*). Most parents don't really think of *uh-oh* as a word at all, let alone an important word. But it is consistently one of the very first words that American and English babies use. As we found out later, Korean and French babies also have their own special words to talk about when they succeed and fail (French one-year-olds who manage to build a rather shaky tower of blocks produce a splendidly Gallic *voilà!*)

Even the children's early names turn out to be more complicated than Augustine, or parents, might think. A father's delight at hearing the baby say his name may fade a bit as the baby hails the dad's best friend with the same jubilant "Dada!" And the mailman. And the TV repairman. It's a slight comfort perhaps to see that the family pet suffers from the same fate: any animal from an anteater to a zebra is a "doggie." One linguist reported that her daughter used *moon* to talk about the actual moon, but also lamps, oranges, and crescent-shaped fingernail parings. Just as the babies extend *uh-oh* or *gone* to new situations, they also extend their early names. They are trying to make sense of the language they hear by applying it to concepts that seem important to them. They use words in a way that makes sense to them, even if grown-ups don't use the words that way.

Initially children use just a few names, mostly for familiar things and people. But when they are still just beginning to talk, many babies will suddenly start naming everything and asking for the names of everything they see. In fact, *what'sat?* is itself often one of the earliest words. An eighteen-month-old baby will go into a triumphant frenzy of pointing and naming: "What'sat! Dog! What'sat! Clock! What'sat juice, spoon, orange, high chair, clock! Clock! Clock!" Often this is the point at which even fondly attentive parents lose track of how many new words the baby has learned. It's as if the baby discovers that everything has a name, and this discovery triggers a kind of naming explosion.

It turns out you can show experimentally that babies at this stage have a new approach to learning words. You can give a baby just one example of a new nonsense word naming a new kind of thing ("Look, a dax!" you say, pointing to an automatic apple corer), and it will become a permanent part of the baby's vocabulary. Weeks or even months later, he'll correctly identify the "dax." Just one salient instance and babies will internalize a word forever (sometimes, of course, with rather embarrassing consequences). The process is called fast mapping. The babies seem to assume at once that the new name they hear names the new object they've just seen. Babies start to fast-map at about the time they have their naming explosion.

Language is as much invented as learned. Babies don't simply soak up associations between names and things or mimic adults' use of words. Instead, they actively restructure language to suit their own purposes. If they need a word for disappearance or failure, they'll happily press *all gone* or *uh-oh* or even *oh bugger* into service. If they need a word for all animals, they'll make *doggie* fit the bill.

If you can make some assumptions about what people are trying to say, that also gives you a substantial leg up in decoding their language. Experiments show that children know something about other people's intentions and use that knowledge to help figure out what words mean. These experiments also show that Augustine was wrong in another way.

If Augustine were right, what would happen if children just happened to be looking at an apple when Dad said, "Where are the pears?" They ought to be stymied. They should mistakenly think that *pears* means *apple.* However, even toddlers don't make this kind of mistake. Suppose you get an eighteen-month-old to look at one new object, for example, a potato masher, while his mother is looking at another object, say a bulb baster. The mother says, "Oh, look, a dax!" Then you put both objects in front of the baby and ask, "Show me the dax." The baby, it turns out, assumes that *dax* means the bulb baster, the thing his mother was looking at, rather than the potato masher he was looking at when he heard the word.

You can set up even more complicated situations like this one. Suppose the toddler and the experimenter sit at a table full of toys and the experimenter picks up each toy and looks at it. Then the experimenter leaves the room, and while he's away another person brings a new toy into the room and leaves it on the table. Now the experimenter returns and cries out, "Oh, look, a dax!" The child assumes that the new object is the dax. Of course, that's what we would assume, too. But when you think about it, that assumption requires a lot of sophisticated knowledge about the other person and about communication. The baby seems to know that people talk about things that are new to them, rather than things that are familiar. And once again Augustine's theory that children learn language by associating a name and a thing turns out to be wrong. In this case the child is looking at many different things when she hears the name, but she connects the name to the thing that is new to the other person.

Putting It Together

Putting words together to make new sentences and more complex meanings is another central part of language. Before they are three, children are working out this part of the language problem as well. Many English-speaking children go through a stage where they start putting words into two-word combinations, like poor Henry's plaintive "Mummy gone!" Two interesting observations suggest that even these very young children already have some idea of grammar. First, they seem to recognize that only some word orders are possible in their language. They say "Mommy gone" but not "gone Mommy"; "more cookie" but not "cookie more." Second, they already use different word orders to express different meanings. "Kiss Teddy" means Mommy should kiss the teddy bear, while "Teddy kiss" means the teddy bear is going to kiss Mommy (undoubtedly assisted by the speaker). These very simple two-word sentences already follow certain rules, even though a two-year-old would never have heard these sentences from anyone else. Just as babies invent meanings, they also invent grammatical rules.

As most English-speaking babies grow older, they start to produce longer and more complicated sentences, but those sentences still sound very different from the sentences of the adult language. We all recognize that two- and three-year-olds have a kind of distinctive "cookie monster" talk (that, of course, is precisely why *Sesame Street*'s Cookie Monster talks that way). What may be less obvious is that "cookie monsterese" is very systematic. Young children systematically leave out word endings, such as the plural *s* or the past tense *ed,* and they omit "grammatical" words such as *the* or *of.* Even if you try to get a toddler to repeat a grown-up sentence word for word, what comes out will be very different. "I don't want the broccoli, I want the cookies" becomes "No want bwocwi, want cookie!" The children have largely made up their own language with its own rules and grammar, just as they decide themselves what the words they use will mean. But the important thing is that they are rules and it is grammar.

Some children, though, especially younger siblings, take quite a different route toward grammar. Rather than starting out with a bunch of individual words and gradually combining them into more complex sentences, these babies seem to take the opposite approach. They seem to get hold of whole sentences and then take them apart into separate words. They start out by grasping the intonation patterns of whole adult sentences, and they babble in a way that mimics those intonation patterns. Often it sounds as if they're quite fluent in a language their parents just don't happen to know, like Klingon or Vulcan. Sometimes startled parents will suddenly hear a whole sentence of English emerge—"I-want-some-cookies-please"—or will hear a few recognizable English words embedded in the Klingon sentence. Eventually and gradually the odd sentences turn into English sentences.

In addition to learning verb and noun endings, babies also have to learn the details of how those verb and noun endings are used. It isn't as straightforward as just adding *s* to a word to indicate a plural. How about *boxes* (with an *iz* sound) and *rods* (with a *z* sound)? And let's not even think about *women* and *children* and *sheep*.

Children learn, and create, systematic rules for dealing with these variations. One of the very first experimental studies of language development demonstrated this. You can show a toddler a picture of an imaginary creature and say, "This is a wug." Now you show him a picture of two of the imaginary creatures and say, "There are two of them, what are they?" By four or five, though not earlier, children will happily say, "They're wugs."

Looking at the children's mistakes, paradoxically, also shows they are learning in an intelligent way. Preschoolers often use invented words like *womans* and *childs* (Alison's sister referred to her large family by saying, "All of we's is childs.") In fact, children often begin by getting some of these words right—a two-year-old may say "children," much to the pleasure and pride of grown-ups, and only later produce the invented form *childs*. The "mistake" actually shows, though, that the toddler has learned the more general rule for making plurals.

Children learning different languages vary even more radically in the ways they approach grammar. We've seen that very young infants are already sensitive to the particular sounds of their particular language. In the same way, even very young toddlers seem to be sensitive to the particular grammar of their particular language.

We just mentioned that English "cookie monsterese" leaves

out noun and verb endings like plurals and past tenses. In fact, English doesn't use those endings very much compared with other languages. Any English speaker who's tried to learn French, Russian, Spanish, Latin, or practically any other language for that matter will remember the countless conjugations and declensions with fear and loathing. (Of course, speakers of other languages find English prepositions and articles equally baffling and repulsive.) Children who are learning other languages pick up and use noun and verb endings much earlier than English-speaking children. Korean-speaking children, for instance, not only use many more early verbs than English speakers, they also use verb endings correctly even when they are using only single words. French-speaking children seem to have almost no trouble picking up the system of grammatical gender, a feat that will seem absolutely astonishing to anyone who has tried to learn French.

How Do They Do It?

How do children manage to do it? There is clearly some genetic foundation that enables human beings to acquire language. That fact has been the focus of much attention in linguistics. But children must also have powerful learning mechanisms, particularly in order to learn the specific properties of their own language. Moreover, grown-ups seem designed to help babies learn.

Word-Blindness: Dyslexia and Dysphasia

Just as there are genetic disorders that make it difficult to understand the mind and the world, there are genetic disorders that make language difficult. And once again, these tragic examples demonstrate that we have an innate endowment that lets us understand and speak.

While all normally developing children effortlessly perform

minor miracles with sound, not all children can do this. There are children who hear perfectly well and who are perfectly intelligent but who still have a hard time with language. We don't know exactly what causes these language disorders, but we do know that the disorders run in families, which points to genetic factors.

Often, language problems become evident only when children start to learn to read and write. Children who are having trouble with the sound system of the language may be able to compensate enough to understand everyday speech. After all, in everyday communication there are lots of cues about what someone is trying to say, including tone, inflection, facial expression, and context. But in order to read and write, you have to translate the system of language sounds directly into a system of written letters. If you don't already have a secure mastery of the sound system, this can be a very tough job. Many dyslexic children, children who have trouble reading and writing, also turn out to have trouble with sounds as well as with written letters. They can't hear the simple distinctions between *r* and *l* or *b* and *p* that most of us could hear at birth. It can be helpful to such children to artificially make the basic sounds of language more distinct. For example, you can use computer programs that alter speech to exaggerate the sound distinctions. Some recent studies suggest that listening to this kind of altered speech improves dyslexic children's reading and writing.

Some children seem to have genetically determined difficulties with other aspects of language. We've seen how normally developing children must master the system of word endings such as *ing* and *ed*. Some people never master this system at all. They may eventually learn word endings but only painfully and one at a time. If you ask them the wugs question ("This is a wug, what are two of them called?"), they fail hopelessly. Their reaction to such linguistic problems is much like the autistic child's reaction to questions about emotion. If you ask them to name a picture of two cats, they may say something like, "I know that . . . *s* . . . *s* . . . more than one is *s*. Cat [pause] *s*." The rest of us would say "cats" without even thinking about it. In some of these cases we not only know there is a family history of these disorders, we can actually trace the site of the defective genes.

Learning Sounds

No matter how rich the genetic basis for language is, babies still have to disentangle all the particularities of Japanese or English or whatever language they are learning. Children clearly must have some powerful abilities to abstract patterns and discover regularities in the language around them. We know less about these learning mechanisms than about what children learn when. But we have some ideas.

Babies growing up in different language environments will hear very different sounds in the speech around them. Pat estimates that by six months of age, the average American baby has heard hundreds of thousands of instances of the vowel *ee* (as in the words *baby, daddy, mommy, cookie*). On the other hand, that same child will have heard hardly any examples of the nasal sound at the end of the French word *non*. We think that babies abstract the prototypical *ee* sound from all these examples. They unconsciously figure out what the ideal *ee* should be like.

Abstracting these mental prototypes has an enormous impact both on how babies hear speech and on the way they coo and babble. The babies unconsciously compare other incoming sounds to the prototypes. If the sound they hear is anything close to the prototype, they simply ignore the differences and assume it was the prototypical sound. So when different people, some of whom may have sore throats or colds, speak to a baby, the baby doesn't actually attend to the distorted sounds they make but "smooths them over" and hears what they meant to say. Babies act as though they heard the prototype. Babies get the speech prototypes from the adults they hear around them, but then they turn around and use the prototypes to decode what the adults are saying, even when they don't speak clearly.

So forming prototypes has great benefits. The downside is that these same prototypes prevent the babies from perceiving what a foreign-language speaker is saying. The babies now hear sounds through the filter of the native-language prototypes. And now the babies' own noises start to sound like the sounds of their particular language. By six to twelve months of age, the baby is no longer a citizen of the world but a culture-bound language specialist, like you and me.

To use our earlier example of the picture of the house, many Americans think the prototypical house has a chimney because that's been true of most of the houses they've seen. Once they form this prototype, it influences the very way they remember chimneyless houses. Presumably people growing up in a place where very few houses have chimneys, such as equatorial Africa, would not make the chimney part of their prototype. And Americans might also have a harder time than a native African at discriminating and remembering African houses.

There are other mechanisms at work in learning the sound system of a language. We mentioned that by the time they're a year and a half, babies raised in Chinese homes sound Chinese and those raised in Swedish homes sound Swedish. We saw that that depends on their understanding how the sounds of their language work. But it also depends on being able to imitate and produce those sounds.

Imitating a sound is a lot more complicated than it seems, however. If you just hear a sound, you don't know what to do with your mouth to produce it yourself. When we hear a Swedish vowel sound like *eu,* we don't really know what to do to make it. Should I raise my tongue or lower it? Should I pucker my lips or not? To make it, you raise your tongue as if to produce English *ee* but pucker your lips as if to produce English *oo* and there you have Swedish *eu.* But if we didn't tell you how to do it, and if you don't speak Swedish or French or one of the other languages that use that sound, you would be clueless. How do babies link the sounds they hear others make to the movements they must make to produce those same sounds?

One idea is that when babies are cooing and babbling, they aren't just exercising their vocal cords and moving their mouths randomly. We believe they are creating a kind of mouth-to-sound map, relating the movements of their speech articulators (their lips, tongue, mouth, and jaw) to the sounds they produce. We know babies play with their arms and legs, moving them to and fro and watching in fascination. In much the same way, they also seem to play with their mouths and listen to the sounds they can produce. Babies will lie in their cribs all by themselves and play with sounds, squealing with delight and producing *ee*'s and *aa*'s and *ba*'s and *ga*'s and even just raspberries for long stretches. By playing in this way, they learn how to make the sounds they hear us produce. They learn that to create a sound like *ee,* they have to raise their tongue, whereas for *ah* they have to lower it.

Moreover, babies aren't just able to imitate us, they are driven to do so. Babies love to copy adult sounds. Pat and Andy found that when five-month-olds listen to a simple vowel like *ee* for fifteen minutes in the laboratory, they will coo back with a vowel of their own that resembles the one they heard.

They can't make perfect *ee*'s, but they already have an idea of what to do with their mouths to make a sound that resembles *ee*. They have learned that *ee* is produced when people raise their tongue and retract their lips. Just hearing a grown-up produce the sound motivates babies to try to produce it themselves.

Remember also that in the last chapter we saw that babies at the same age do something akin to lipreading. They prefer to look at the face of a person mouthing a vowel that matches one they are listening to, rather than at a face mouthing a different vowel that doesn't match. This is another sign that they are linking up the sounds they hear with the mouth movements that make those sounds. This combination of abstracting prototypes, playing with sounds, and imitating sounds seems to help children break the speech code.

Learning How to Mean

Why do babies use odd words like *gone* and *uh-oh,* and why do they start to fast-map names? We saw in the last chapter that when babies are about eighteen months old, just when they're learning to talk, they are also learning a great deal about the way objects can appear and disappear, about how they can use tools, and about how objects fit into categories, and they're fascinated by all these problems. Alison suspected that these changes in the way children solve problems might actually be connected to their early words. A fascination with disappearance, and not the rituals of cereal eating, could account for the otherwise mysterious prevalence of *gone.*

How could you test the idea that these weird early words were the result of the problems children were trying to solve? You can visit a baby every few weeks and give him different kinds of problems to solve. At the same time, you keep track of his new words. Alison and Andy found out that babies start to use *gone* within a week or two of the time they first solve the hardest keys-under-the-washcloth hiding problem, sometimes a little before and sometimes a little after. The word for disappearance and the concept of disappearance seem to emerge together.

Just as *gone* is related to object disappearances, *uh-oh* is related to children's ability to use tools. Remember in the last chapter we talked about how babies learned to use a rake to get a faraway toy. Babies worked out how to use tools like that within a few weeks of the time they started to use words such as *uh-oh,* just as they solved disappearance problems at about the time that they first used *gone.*

Naming turns out to be connected to understanding a rather different aspect of the world. We saw in the last chapter that children learn about how objects fit into categories at about this age. They start to divide mixed-up groups of objects into several different piles, with a different kind of thing in each pile. Children start to do this at about the same time they begin to fast-map new words and to use lots of new names. When we visited babies once every few weeks, we found that just when they suddenly used a lot of new names, they also started to sort mixed-up objects in a new way. Babies "get" the idea that everything has a name and that everything belongs in a category at the same time. So early words often appear at the same time children are solving relevant new problems.

What's going on here? We think it's a bit like college, really. Think about the first time you learned about some new concept in a class like introductory physics. If you were really interested in the course, and not just in getting into medical school, you went to lectures and read the textbook, trying to understand how physics worked. As you studied, you came across peculiar new words such as *entropy* that at first you understood only vaguely. On the other hand, you could see that they were relevant somehow to the physics problems you were trying to work out. Entropy had something to do with heat loss and something to do with disorder, though you couldn't quite tell what being cold and disorderly had in common (aside from being characteristic of your dorm room, not to mention your boyfriend). Then one day there was a magic moment when everything clicked; you "got" entropy. Part of getting it was really understanding the word and really being able to use it convincingly on the exam, but part of it also was really getting the idea, really getting the concept that brings heat and chaos together and being able to solve problems that require it. ("Briefly define . . ." and "Solve, show your work" both are likely to appear on the exam.) You probably still didn't entirely get it, though, and you undoubtedly used the word in peculiar ways that revealed your continued ignorance to the godlike and omniscient TA's.

Baby Henry's *gone* seems about like the freshman's *entropy,* but without test anxiety interfering. Henry was working on these baffling problems of appearance and disappearance and kept hearing the people around him say *gone* just as some peculiar disappearance took place. One fine day he "got" both the word and the concept of disappearance.

So babies' guesses about the meaning of their first words are informed by the other kinds of cognitive progress they've already made in infancy. Their ability to solve the Language problem is closely tied to the particular ideas they've already developed in solving the External World problem. The mechanisms that drive children to make coherent sense of the world also lead them to pay attention to the words they hear and to learn how to use those words themselves.

"Motherese"

Grown-ups are the third component in the solution to the Language problem. We've mentioned that we sound positively silly when we talk to babies. If you listen to mothers talking to other grown-ups and then to their babies, you hear a strange shift in their voices. A mother says to a friend, "The traffic, it was awful, and I had to park and there was a delay and I didn't have change for the meter . . . ," droning on about the events of the day. Then, with hardly a pause, she turns to the baby in her arms and coos, "Hiiiiii, sweeeeetie. How's my baaaaby?" She swoops in with her voice and face. "Ooooh [tickling the baby's cheek], open up thooose eeeyes. Ooooh, you're sooooo cute. Can Mommy have a big, big smile? Mmmmm, give me big blue eeeeyes, toooo!"

Anyone listening to a parent talking to a baby knows that this is definitely not a job-interview voice. It's the voice of a playful, animated, warm, and practically giddy person totally absorbed in the little bundle in front of him or her. Out of context it sounds ridiculous. But put us in front of a baby and we all do it, mothers, fathers, grandparents, friends—even four-year-olds speak motherese to their baby brothers and sisters. (We have occasionally heard a macho, deep-voiced dad say, "I never talk baby talk, don't believe in it," and in the next breath turn to the baby and say, in a voice a full octave higher, "Dooo I, sweetie? No, I dooon't talk baby talk to YOOOOU.")

And babies love it. When you give babies a choice of what to listen to, a kind of baby Nielsen rating, they choose to listen to mothers talking to infants over mothers talking to other adults. In these tests the babies sit in an infant seat, and slight turns of the head give them eight-second sound bites of either a mother talking to a baby or that same mother talking to another adult. Babies get to choose which tape to listen to simply by turning their heads in one direction or the other.

The tests show that babies' preferences have nothing to do with the actual words mothers use. Babies choose motherese (or "parentese" or "caretakerese") even when the speaker is talking in a foreign language so infants can't understand the words, or when the words have been filtered out using computer techniques and only the pitch of the voice remains. Apparently they choose motherese not just because it's how their mother talks but because they like the way it sounds. Motherese is a sort of comfort language; it's like aural macaroni and cheese. Even grown-ups like it. Pat's graduate students discovered that listening to the lab tapes of motherese in a foreign language was a wonderful therapy for end-of-term stress. The mother's voice is an acoustic hook for the babies. It captures babies' attention and focuses it on the person who is talking to them.

The elaborate techniques of computer voice analysis reveal exactly what it is we do when we talk to an infant. The pitch of our voice rises dramatically, sometimes by more than an octave; our intonation becomes very melodic and singsongy; and our speech slows down and has exaggerated, lengthened vowels.

Motherese is a universal language. People across all cultures do it when they talk to their infants, even though they usually aren't aware of doing it at all. When mothers listen to recordings of themselves producing motherese, the reaction is: That can't be me. I sound really stupid. Should I be doing that? But they do it intuitively, without conscious awareness.

Why do we do it? Do we produce motherese simply to get the babies' attention? (It certainly does that.) Do we do it just to convey affection and comfort? Or does motherese have a more focused purpose? It turns out that motherese is more than just a sweet siren song we use to draw our babies to us. Motherese seems to actually help babies solve the Language problem.

Motherese sentences are shorter and simpler than sentences directed at adults. Moreover, grown-ups speaking to babies often repeat the same thing over and over with slight variations. ("You are a pretty girl, aren't you? Aren't you a pretty girl? Pretty, pretty girl.") These characteristics of motherese may help children to figure out the words and grammar of their language.

But the clearest evidence that motherese helps babies learn comes from studies of the sounds of motherese. Recent studies show that the well-formed, elongated consonants and vowels of motherese are particularly clear examples of speech sounds. Mothers and other caregivers are teachers as well as lovers. Completely unconsciously they produce sounds more clearly and pronounce them more accurately when they talk to babies than when they talk to other adults. When mothers say the word *bead* to an adult, it's produced in a fraction of a second and it's a bit sloppy. But when mothers say that same word to their infants, it becomes *beeeeeed,* a well-produced, clearly articulated word. This makes it easier for infants to map the sounds we use in language.

In fact, adults may even adapt motherese to suit the characteristics of their particular language. Pat recently discovered that Swedish, Russian, and English mothers each make subtle, unconscious variations in the way they talk that are tailored to the particular language they use. Swedish motherese makes the vowels of Swedish sound much clearer than does ordinary Swedish speech to adults. Similarly, English motherese seems particularly well designed to make the vowels of English sound clear. The Swedish and English mothers provide the babies with just the range and variety of sounds they need to abstract the right prototypes for each language. This is a particularly important result because it makes it especially likely that babies are taking advantage of motherese to learn the sounds of their language. If motherese were no more than a universally attractive and comforting set of sounds, it might not play much of a role in the details of what babies learn. But in addition it is exquisitely adapted to help babies solve the particular problems of their particular language. That makes us think it must be having a real effect.

Studying babies leads us to realize that, however effortless and instinctive our adult ability to speak may seem, it is actually the outcome of a great deal of learning. There is nothing contradictory about saying this and saying that language also has an important innate component. In fact, the point is not that language is the product of both nature and nurture, innate knowledge and learning. Rather, nature and nurture are inseparably intertwined. The innate endowments enable babies to use their powerful learning mechanisms to take advantage of the information they receive from grown-ups. The fact that babies can already make the right distinctions between sounds at birth enables them to reorganize and reshape those distinctions in light of what they hear their parents say. The fact that babies already organize their world, and are motivated to make sense of it in new ways, also motivates them to learn new words and shapes the meanings they give those words.

Linguists sometimes use the term *bootstrapping* to describe this process. Babies take what they already know and use this as a basis to learn more: they pull themselves up by their own bootstraps. Although language learning is different from scientific-theory formation in many ways, both kinds of learning involve this sort of bootstrapping: scientists also use their current theory as a basis to formulate new theories. As anyone who has actually helped a young child pull up real boots knows, a few tugs and nudges from grown-ups come in handy, too.

CHAPTER FIVE

What Scientists Have Learned About Children's Minds

So far we've been talking in detail about what children know about a vast array of different topics, from broccoli preferences to toy-car trajectories to the difference between *p* and *b*. We've seen that in three brief years there are enormous changes. What newborns know is very different from what one-year-olds know, which is different in turn from what three-year-olds know. We've also seen that children tackle profound and significant problems. They learn that other people have minds, that the world exists independent of their subjective experience, and that words have meaning. These are *hard* problems. In each case there is a gap—in fact, a yawning chasm—between the data that enters the children's eyes and ears, the light and sound waves, and the conclusions children reach about people and the world and language.

We know more about what children learn than about how they learn it. The mechanisms of learning may be quite different for different problems. We've already seen that understanding how words sound is very different from understanding how objects move. We need to develop detailed, specific theories that define what children know at each point and how they learn more. That will mean hard scientific work. And no single scientist can attack more than a tiny piece of the problem. Whole careers may be devoted to understanding just what six-month-olds know about sounds or what one-year-olds know about objects.

Still, looking across all the different topics, ages, and specific theories, you can see some common basic ideas. There would certainly be arguments about the details. And other theorists might well reject one or the other of our proposals about how children learn. But a big picture does seem to emerge from all the scientific particularities.

Evolution's Programs

The first basic idea is that babies can solve the ancient problems because the human brain is like a biological computer designed by evolution. This idea is the best way we currently have of bringing together what look like two completely different things: our knowledge of the world and the three pounds of gray jelly in our skulls. How could a physical object, a brain, also be something that thinks or reasons or knows things?

The old answer to that question, Descartes's answer, for example, was that there were two fundamentally different kinds of stuff in the world: physical stuff such as rocks, trees, and bodies, and mental stuff such as souls, spirits, and minds. Human beings were a combination of a physical body and an incorporeal soul. The philosopher Gilbert Ryle described this idea as "the ghost in the machine."

That answer is incompatible with all we have learned from five hundred years of science. From the perspective of science the mind that is composing these words must be just as much part of the physical world as the wooden desk in front of it. So what is it that makes some physical objects, like developmental psychologists, able to think, reason, and know, while other physical objects, like wooden desks, are unable to do those things?

The computer this book is being written on lies between the mind and the desk, figuratively as well as literally. The computer is a physical thing, like the desk, but it has some abilities that are more like mental abilities. It can alphabetize a list, make an index, play chess, or solve a complex statistical equation. If it were connected to the really powerful expert-system computers at the university, it could diagnose an illness or analyze the rocks on Mars.

How can the computer do this? The computer on the desk is just a few pounds of silicon and plastic, not much more impressive than the few pounds of carbon and water in our skulls. Forty years ago it would have been many more pounds of vacuum tubes; forty years from now it may be a set of quantum fluctuations in subatomic particles. What's important is not the stuff the computer is made of but the way that stuff is organized—the way it works.

The computer can index and play chess and solve equations because of the program the computer runs. The program determines what the computer does. If I know this computer is running Microsoft Word, for instance, I know something about what the computer can do. In fact, I know much more than I would if I simply knew what kinds of circuits and chips this computer uses. Even a computer with a very different physical structure could be a computer running Microsoft Word and could therefore do the same sorts of things as this computer.

Programs ultimately do two things. A computer program takes in information and translates it into a string of symbols. Then it has rules for manipulating and rearranging those symbols. Cognitive scientists often talk about those internal symbols as "representations." The word-processing program manipulates symbols that represent the words and sentences of this book.

The program finds out about those sentences in a very simple way: you type a string of letters on the keyboard. That's the input to the computer. The computer program translates that information into a set of more abstract symbols, the sort of symbols that are the "words" of a programming language. The program might label some of the strings of letters you type as the equivalent of "Words that start with A" and label others as "Words that start with B." Those symbols are used in the rules that tell this program, for example, to list words in alphabetical order. These rules in the programming language might say the equivalent of "Put words that start with A before words that start with B." The program then systematically manipulates and rearranges and rewrites those more abstract symbols. Those rules allow the program to alphabetize your words, index them, correct them if they're misspelled, even format your entire document. The new strings of symbols—the representations—it comes up with at the end of this process are responsible for the output of the computer. The computer now systematically translates those abstract symbols into patterns of letters on the monitor. The output of the program may be quite different from the input it started with. After the program has done all its alphabetizing and indexing and spell-checking and formatting, the words and sentences that finally appear on your screen will be different from what you originally typed in.

Of course, the word-processing program uses very simple representations and rules. Ultimately, though, even the supercomputers on campus do versions of the same thing. The remarkable discovery that led to the invention of computers is that this process of translating and rearranging symbols can be done automatically, by a physical system. There is a further process that translates the abstract symbols of the programming language into a set of very specific instructions for switching the circuits in the computer. The programming language eventually is translated into what computer scientists call the machine language, and that leads to a particular pattern of physical events in the computer. The remarkable discovery that followed was that a physical system like this could do many things we once thought only people could do.

The basic idea of cognitive science is that we can think and reason and know because our brains run very powerful programs—programs that are much more powerful than the programs even supercomputers can run now. Our brains take in the input of the light waves and sound waves entering our eyes and ears and systematically transform and rearrange it. They turn that information into more abstract representations and use rules to transform those representations. Eventually those transformed representations are responsible for the output—what we experience, say, and do.

The thing that makes the developmental psychologist different from the wooden desk isn't so much that the psychologist is made of cells and the desk is made of wood. It's that the cells that make up our brains, like the electronic circuits in the computer, are organized to work in this special way. It's rather like the basic *Scooby-Doo* plot: the ghosts in the machine turned out to be ordinary physical matter in disguise.

A special kind of program is particularly relevant to the ancient problems of knowledge. In principle, programs can take all kinds of input and turn it into all kinds of output. They can take a random list of words and turn it into an alphabetized list, or they can take an opponent's chess move as input and come up with an answering move as output. But some programs have a particular, special relationship between inputs and outputs. There are programs that are designed to take a video image and translate it into a description of objects. Similarly, there are programs that are designed to take a list of symptoms and diagnose an illness. Others take the electromagnetic spectra collected by a Mars rover and figure out which minerals are in the rocks. These programs all start out with input that comes from some real object or phenomenon in the world: the pixels on a video image of an object, a list of symptoms, the spectra of the light reflected from the rocks. Then they are designed to generate representations that will be a more accurate picture of the objects than the input itself would be. The representations specify objects, or illnesses, or the mineral composition of the rock.

If the system gets these representations right, then it can generate output that includes accurate predictions. For example, if a computer vision system gets the right representations of objects, it can predict what will happen if the objects move. If a diagnostic program gets the illness right, it can predict what will happen if you undertake a particular treatment. If a geology program gets the mineral composition of the rocks right, it can predict what will happen if you apply a particular chemical test to the rocks.

Programs like these figure out what the world is like. They try to solve the problem of knowledge. One philosopher-turned-computer-scientist calls designing these programs android epistemology. Programs like this are very complex and difficult to design. And yet computer vision is still not even in the same ballpark as human vision, and programs that diagnose illnesses and analyze rocks are still not as skillful as doctors and geologists. Such programs have been designed, however, and they are very powerful and useful.

Cognitive scientists think that people have programs like

these, programs that are even more powerful than the existing computer programs. Our brains were designed by evolution to develop representations from input that accurately approximate real things in the world. Those programs give us the same advantages they give our computers: they let us predict what the world will be like and so act on it effectively. They are nature's way of solving the problem of knowledge.

The *Star Trek* Archaeologists

The job of computer scientists, of course, is to design the programs that let electronic computers accomplish those impressive feats of thinking and knowing. The computer scientists have to figure out how to make programs that get to the right kind of output from the right kind of input. But our job as cognitive psychologists is rather different and even harder. We are more like archaeologists than engineers.

Actually, it's a familiar *Star Trek* story. We have landed on a planet that already contains amazing biological computational devices. They were designed eons ago over millions of years by a force far more powerful than any we possess. The one thing we know about them for sure is that they employ incredibly advanced technology. There are no operating manuals, no wiring diagrams, no *Homo Sapiens for Dummies.* We can't even hope that sometime in the last few minutes of the show the all-powerful designing intelligence will take over one of our crew and explain its intentions in a suitably resonant and spooky voice (the usual *Star Trek* resource in these situations). We're on our own.

What can we do to figure out how these devices work? Well, one thing we could do is try splitting them open and looking inside, especially if we could do that without breaking them. As we'll see in the next chapter, that's what neuroscientists have started to do. But another thing we could do is try to figure out what program they are running. We could put in input and see what kind of output comes out, type things on the keyboard and see what appears on the monitor. If we were to do that cleverly enough, we could figure out, at least in general terms, what kind of programs these devices run and what kinds of representations and rules their programs use.

This isn't just a utopian sci-fi project. There are some aspects of the biological computational devices on this planet that we're close to understanding in this way, even in detail. For instance, we are close to understanding some parts of the programs that let us transform sound waves into words and light waves into images of objects. We are further from understanding other aspects of these devices. We have some general ideas about how the biological computer programs understand the nature of people and things and how they use meaningful words, but nothing like details. We don't even have a clue about other aspects of these devices. We basically don't know anything, for example, about how or why the biological computers on our planet are conscious. We know that they generate what philosophers call phenomenology—the special subjective "feel" of our conscious experiences, the particular way a color looks or a noise sounds to us. But we have no idea how they do it.

What does all this have to do with babies and young children? People thought for millennia that babies and young children couldn't think, reason, or know. This is partly because of the way we conceive of minds in everyday life. In everyday life we tend to be quite sure that we have minds and reasonably sure that other folks who are like us, particularly folks who talk like us, have minds. If we want to find out what (or if) someone thinks, it's easy: we strike up a conversation.

We don't remember much about what we were like before

we were three years old, and we remember nothing about what we were like as babies. It can be hard to carry on a conversation with very young children, and babies, of course, can't talk at all. So it might seem reasonable to conclude that babies and young children don't think and, in fact, that we ourselves didn't think when we were that young. (Even some modern philosophers have, quite seriously, made just this claim on the basis of this kind of evidence.)

When we start to treat thinking as a kind of computation, though, our criterion for distinguishing between thinking and nonthinking creatures turns out to be rather different. If a computer can think, or at least can play chess and solve equations and diagnose illnesses and analyze rocks, then a baby, whose brain is infinitely more complex than the most sophisticated computer, might be able to think, too. If a machine can run a sophisticated program, then a baby might be able to, as well. To find out, we have to see if there are systematic relations between the baby's input and the baby's output—between what goes in at the keyboard and what shows up on the monitor.

From the perspective of the *Star Trek* archaeologist there's no reason that the big computational devices who look more like us and have more user-friendly interfaces should be fundamentally different from the little ones. If cognitive psychologists are clever enough about giving babies the right kind of input, and about interpreting their output, we should be able to work out their program, too. The last three chapters of this book, and the last thirty years of developmental psychology, really have been about doing just that. We've developed ingenious ways to give babies and young children the right kind of input and to interpret their output. We show them a moving face and see if they produce a matching facial expression.

We play them a tape of *p* and *b,* and see if they turn their heads. We ask them a simple question—"What does Nicky think is in the box?"—and see if they say pencils or candy.

In fact, as *Star Trek* archaeologists, we might be better off tackling the little devices rather than the big devices. One odd and interesting thing we know about these machines is that all the big ones start out small. The little machines actually turn into the big ones. If we want to understand the basic mechanisms that make these devices tick, perhaps we should start out small, too.

So what can we say about the programs babies run? What kinds of representations and rules do their programs use? We still are far from knowing the details of how these systems work. One important reason is that the babies must work in ways that are very different from the way existing man-made computers work. But we have learned enough to know what the broad outlines of the babies' programs must be like.

We'll summarize this big picture by elaborating on the three ideas we've presented in previous chapters.

Foundations. Babies begin by translating information from the world into rich, complex, abstract, coherent representations. Those representations allow babies to interpret their experience in particular ways and to make predictions about new events. Babies are born with powerful programs already booted up and ready to run.

Learning. Their experiences lead babies and young children to enrich, modify, revise, reshape, reorganize, and sometimes replace their initial representations, and so to end up with other, quite different rich, complex, abstract, coherent representations. As children take in more input from the world, their rules for translating, manipulating, and rearranging that input also change. Rather than having a single program, they have a succession of progressively more powerful and accurate programs. Children themselves play an active role in this process by exploring and experimenting. Children reprogram themselves.

Other people. Other people, especially the people who take care of children, naturally act in ways that promote and influence the changes in the children's representations and rules. Mostly they do this quite unconsciously. Other people are programmed to help children reprogram themselves.

Foundations

The most striking new fact to come out of thirty years of developmental research is just how much even very young babies know. Twenty-five years ago, when we were in college, we still heard respected psychologists proclaim that newborn babies had no cortex, that they had only the simplest automatic responses, that they were, in fact, slightly animate vegetables—carrots that could cry. Piaget himself thought that newborn babies had only reflexes.

Not surprisingly, there was a sense that such simple creatures were hardly worth studying: infancy was barely even an academic field until the 1970s. At the first meeting of the International Conference on Infant Studies in 1978, the researchers fit into a small hotel conference room. At first, every new discovery about young babies, no matter how methodologically rigorous, was greeted by a kind of profound disbelief that seemed to go beyond the usual scientific reluctance to accept new discoveries. It was as if the very idea that babies could think and believe, learn and know, was deeply unacceptable.

Of course, all the historical factors that led philosophers to be skeptical about children's minds apply in spades to young babies. Very young babies don't seem to do very much, and exploring their minds required particularly clever new techniques and videotape technology. Moreover, very young babies were, at the same time, familiar and even ubiquitous and yet peculiarly invisible. They were part of the everyday domestic life that academics of all stripes, from scientists to sociologists to historians, thought was beneath notice.

As more women became scientists and more male scientists began to take care of young babies, and as videotape technology became available, we began to pay more real attention to babies. That in itself made the "crying carrot" picture look a lot less likely. People who take care of young babies usually believe that babies can think, but it was easy, at first, for scientists to dismiss those intuitions (they were, after all, literally old wives' tales). It got a lot harder, though, when the scientist and the caregiver were the same person, and when you could back up your intuitions with videotaped proof. Old wives (and one old husband) are writing this book.

Whatever the larger historical influences, the scientific battle was hard-fought. But by now it largely has been won. The sociological zeitgeist may have contributed to the victory, but the real weapons were the familiar scientific ones: careful and ingenious experiments, replications across laboratories, good arguments, and the conversion of the next generation. The International Conference on Infant Studies books big convention centers for its meetings now, and there are any number of academic journals and associations devoted to infancy (somewhat mixed blessings, perhaps). Thoughtful-looking babies have even made it onto the covers of *Time* and *Newsweek* and the science page of *The New York Times.* Today very few scientists would say that babies are born with only a few reflexes and fixed responses to stimuli.

Even the youngest babies seem to have representations of the world. They have symbols inside their minds that represent the world outside, in much the way the symbols of computer programs do. They take in input from the world, the light waves and sound waves, and they have rules that transform that input into very different kinds of representations. Those representations are responsible for the output: the babies' expressions, gestures, and actions.

But even computer programs have many different kinds of representations and rules. They range from the simple representations and rules in a pocket calculator to the very complex and abstract representations and rules in the programs that diagnose illnesses or analyze Mars rocks. What are the babies' representations and rules like?

First, the babies' representations are rich and complex. As we've seen, they include ideas about how their face resembles the faces of others, how objects move, and how the sounds of a language are divided. The young babies' world is not simple. Babies translate the input at their eyes and ears into a world full of people with animated, expressive faces and captivating, intricate, rhythmic voices. It's also a world full of objects with complex multidimensional structure that move in a dizzying variety of ways.

Babies' representations are also abstract. They go beyond the data of immediate sensation. Most obviously these early representations link information from different senses: they link the way the tongue feels and the way it looks, the bounce of a ball and the *boing* sound it makes, the look of an open mouth and the sound of an *aah*.

But the representations go beyond sensation in other, more profound ways. They turn facial expressions into emotions. They convert two-dimensional images into three-dimensional objects. They take a continuous stream of noise and divide it up into discrete speech sounds. Even newborn babies end up with representations that are radically different from the input at their eyes and ears. The babies' world isn't concrete any more than it's simple. Babies already see the soul beneath the skin and hear the feeling behind the words.

These representations and rules lead young babies to interpret what happens to them in particular ways—to pay attention to some things and ignore others. At first they are particularly captivated by faces and voices; within a few days they pay special attention to familiar faces and voices. At first they pay special attention to the way things move and less attention to their shape or color or texture; later they will start to pay more attention to these properties of objects. At first young babies pay attention only to some changes in sounds and not others; later they will no longer attend to sound changes that once intrigued them.

Finally, the babies' representations and rules allow babies to form expectations, and even to make predictions, about new things that will happen in the future. When the babies' program gets information about a current event, it can generate a representation of a future event. When young babies see a toy car go behind the screen, they look ahead to the far edge of the screen, expecting it to appear there. When young babies flirt, they expect that their coos will be answered by adult goos. When they see an open mouth, they expect that they will hear an *aah* sound. They react in characteristic ways when their predictions turn out to be wrong and their expectations are dashed. They show conflict when the toy car doesn't appear to behave as it should, and they are distressed when their flirtatious advances are met with an impassive stony face. Just as the babies' world isn't simple or concrete, it also isn't limited to the here and now. Even very young babies can remember what happened in the past and predict what will happen in the future.

The significance of this inborn program goes beyond just the simple fact that there is a lot there to begin with. The baffling problem for philosophers and psychologists was always how we get from the raw, undigested matter of sensation—the "blooming, buzzing confusion"—to an understanding of the world. How do we even know which kinds of sensations to pay attention to? The answer the babies give is that we are never dealing with raw matter. There never is a blooming, buzzing confusion. From the very beginning we can understand the world, pick and choose what's important, know what to expect. From the time we're born, we run a program that translates the light and sound waves into people, objects, and language.

Learning

It is a sad truth in science as well as politics that one generation's revolution becomes the next generation's orthodoxy. Now that the idea that babies know a lot to begin with has become almost universally accepted, the idea that they also don't know a lot, and that they learn a great deal, suddenly seems radical. There are many controversies about how, and how much, babies learn.

When it comes to learning, the biological computers look very different from man-made computers. The initial program of very young babies is amazingly sophisticated, especially when you think that all this software power is encased in such a helpless package. We've seen that many of the things new babies can do are far beyond the capabilities of current computers, in fact far beyond the capabilities of the most advanced epistemological androids. But the even more amazing thing is that that initial program seems to spontaneously turn into an even more powerful and accurate program.

The three-month-old's program is apparently quite different from the one-year-old's program or the four-year-old's program. When we put the same input into a three-month-old and a four-year-old, we get very different output. Type the same things into the keyboard, and very different messages will appear on the monitor. When the Japanese three-month-old is presented with *r* gradually changing into *l*, he hears two distinct sounds, while the Japanese one-year-old hears only the same sound. The fourteen-month-old sees you look disgusted with the Goldfish crackers and pleased with the broccoli and gives you the Goldfish anyway; the eighteen-month-old gives you the broccoli. A three-year-old sees the deceptive candy box full of pencils and says that Nicky will think there are pencils in the box. The four-year-old says Nicky will incorrectly think there is candy in the box.

When a three-month-old, a one-year-old, and a four-year-old look at the same event, they seem to have very different thoughts about it. They seem to transform the light waves and sound waves into different representations, and they use different rules to manipulate those representations. Children don't have just a single, fixed program that gets from input to output. Instead, they seem to switch spontaneously from using one program to using another, more powerful program. That makes babies and children look very different from the computers we have now. And it makes our job as *Star Trek* archaeologists much more difficult.

How can we explain these changes? One idea might be that the changes are simply a result of the fact that babies grow, the way caterpillars change into butterflies as they grow, or the way we develop breasts and beards as we grow and reach puberty. The changes might just involve a genetic blueprint that unfolds on a particular maturational timetable. The child's program for understanding false beliefs might appear when she's four the same way her breasts appear when she's twelve. After all, we don't think that the caterpillar learns how to be a butterfly. Similarly, we might not think that the child learns about false belief any more than she learns how to have breasts.

Another very different possibility is that we change our ideas about the world just by taking in more and more information about it. We simply accumulate more and more input. Then we associate some pieces of that input with other pieces. We hear the dinner bell and food comes, and after a while we link the bell with the food. We give a particular answer to the experimenter's question and we get praise, and after a while we try to give answers like that. Babies could end up linking particular inputs to each other and to particular outputs in this sort of specific, piecemeal way.

The Developmental View: Sailing in Ulysses' Boat

We think that neither of these pictures gets the facts of development right. There probably are parts of the babies' program that mature, and babies probably do learn some things by detecting associations in the input and associating inputs with outputs. But that can't be the whole story or even most of the story. Most of us who have sat face-to-face with babies and young children for a long time find the caterpillar analogy pretty implausible, but we find the dinner bell analogy pretty implausible, too. Within developmental psychology there are many different theories about how babies' understanding of the world changes, with many different ideas about the balance between maturation and experience.

Our own view is that children's whole conception of people, objects, and words changes radically in the first three years of life. And it changes because of what children find out about the world. We already said that babies start out with complex, abstract, coherent representations of the world and rules for manipulating them. They use those representations and rules to make sense of their experience. And they also use them to make predictions about what the world will be like. But once babies have done this, they can compare what they experience with what they predicted. When there are discrepancies, they can modify their representations and rules. When they see a new pattern in their experiences, they can create new representations and rules to capture that pattern. Often, babies seem to change a lot of representations and rules at once, rather abruptly. The new representations and rules lead to new experiences and predictions, and the process of creating and testing ideas starts over again. What we experience interacts with what we already know about the world to produce new knowledge, which enables us to have new experiences and to make and test new predictions, which enables us to produce further knowledge, and so on.

The philosopher Otto Neurath compared knowledge to a boat we rebuild as we sail in it. To keep afloat during his thirty years of wandering, Ulysses had to constantly repair and rebuild the boat he lived in. Each new storm or calm meant an alteration in the design. By the end of the journey hardly anything remained of the original vessel. That is an apt metaphor for our view of cognitive development. We begin with many beliefs about the world, and those beliefs allow us to understand what's going on around us and to act—they let us navigate our way around. But as we do, we get new information that makes us change our beliefs and therefore understand and act in new ways.

We see this sort of change in many different areas of children's development. Babies start out linking their own internal feelings to the expressions of other people. That link lets them imitate and flirt, and puts babies and parents in that delightful, intimate, romantic cocoon. That initial representation, though, changes as children come to understand more about the world. By the time they are eighteen months old, they've constructed a more complicated and rather different picture that integrates people and objects. They understand something about how people differ as well as understanding how they are similar. That changes them from the geniuses of intimacy to more complicated creatures who can be both monsters of perversity and angels of empathy. By three or four they revise those representations once more. They start explaining what people do in terms of what those people think about the world, and they discover that different people may think different things about the world. That discovery, in turn, gives children new abilities to deceive and to be skeptical—but also to truly understand another person's perspective.

Babies start out believing that there are profound similarities between their own mind and the minds of others. That belief gives them a jump start in solving the Other Minds problem. But during the first three years they also observe the differences in what people do and say. Those differences stem from the fact that all minds aren't actually entirely alike. Babies and young children watch and listen with careful focused interest as their mother refuses to let them touch the lamp cord or as their older brother tells them they are completely wrong. This new evidence makes babies revise the beliefs they started out with.

Similarly, babies start out knowing that space is three-dimensional and that objects move in predictable ways. They even reach out to objects and shrink away from them. By the time they are eighteen months old, as they watch and manipulate the things around them, as they play peekaboo and sort things into piles, they see those objects act in new ways and they look for ways to explain what they see. They learn that three-dimensional moving objects continue to exist no matter how they appear or disappear, and they learn that all those objects belong in categories. By three or four they have transformed those first categories into biological species and "natural kinds," as they begin to understand that kittens become cats and that tigers have guts inside and rocks don't.

Finally, babies start out making all the possible distinctions between the sounds of languages. Like citizens of the world, American newborns can distinguish African Kikuyu sounds as well as English sounds. By twelve months, as they repeatedly hear the sounds of their own language, babies create new representations that reflect the sound categories of their particular language. One-year-old American babies can't discriminate Kikuyu categories anymore, but they can discriminate the English categories better, and they have even become "English-sounding" babblers.

In each case the things babies already think influence where they will go next. They determine which events will engage them, which problems they will tackle, which experiments they will do, even which words they will listen to. Then babies change what they think in the light of what they learn.

Babies have another ability that man-made computers lack. They can do things. They can actively intervene in the world as well as passively learn about it. A one-year-old can reach for a new rubber duck, put it in his mouth, bang it against the side of the tub, splash it in the water, and watch his father's reactions to all of this. A key aspect of our developmental picture is that babies are actively engaged in looking for patterns in what is going on around them, in testing hypotheses, and in seeking explanations. They aren't just amorphous blobs that are stamped by evolution or shaped by their environment or molded by adults.

In Chapter Three we described how children *need* to figure out what's going on around them—they have a kind of explanatory drive. This drive pushes them to act in ways that will get them the information they need; it leads them to explore and experiment. The apparently pointless activities we call play often seem to be the result of this drive. Babies who are figuring out what people think play imitation games; babies who are figuring out how we see objects play hide-and-seek; babies who are figuring out the sounds of language babble. It's all very serious fun.

Our archaeological investigations of children's programs tell us that the biological *Star Trek* computers on this planet work very differently from the man-made computers we have now. If we were going to try to reverse-engineer the biological computers, we would have to give our computer scientists a very demanding set of specifications.

The baby computers start out with a specific program for translating the input they get into accurate representations of the world and then into predictions and actions. But the interesting thing about these computers is that they don't stop there. Instead, they reprogram themselves. They actively intervene in the world to gather more input and check their predictions against that new input. The things they find out lead them to construct new and quite different representations and new and quite different rules for getting from inputs to representations. If we wanted to make a new computer as powerful as the biological computers, this is what it would have to be like. (Though, of course, if all we wanted to do was make another computer that could do all this, we already know how to do it, and it's a whole lot easier and more fun than programming.)

Big Babies

If we're right that this is how babies and young children work, then maybe we adults work this way, too. We are, after all, just big, old babies. We may also be sailing in Ulysses' boat, taking off from what we already know, gathering more information about the world, and revising our views in the light of what we find out.

In fact, this general developmental view of learning may apply quite broadly. We have already seen that it applies to learning about language as well as learning about the world. It applies to kinds of learning that we might think of as more perceptual, for example, learning speech discriminations, as well as those we might think of as more conceptual, such as learning about objects. It applies to learning that is almost completely unconscious, as is much language learning, as well as learning that seems much more consciously accessible, such as learning about other minds.

The general developmental view may also apply quite broadly to adults. Making pictures or poems or even music is like this, too. Artists also create complex, abstract, coherent representations of the world (even in nonrepresentational art). Those representations go beyond what we see to capture something we think is true. Those artistic representations spring from the work that has already been created, but they also extend and revise artistic traditions and introduce new methods to solve new problems. Artists actively experiment with new possibilities and change what they do in the light of what they find. And, of course, a new artistic achievement changes the way we see the world, sometimes quite literally.

It seems at least possible that the developmental view could also apply to the way we make moral and political decisions. We take off from some basic ideas about how people should treat one another or how a society should be organized. But then we also experiment with new ways of thinking about people and organizing societies, ways we believe may work better. We observe what conditions lead people and societies to thrive or degenerate. We revise our ideas in the light of new things we learn, and particularly in the light of the results of our practical political experiments.

The Scientist as Child: The Theory Theory

Human learning, then, may often be like Ulysses' boat. But one specific kind of adult learning seems *especially* like children's learning. A number of developmental psychologists have argued recently that what children do looks strikingly like what adult scientists do. Children create and revise theories in the same way that scientists create and revise theories. This idea seems to explain at least some types of cognitive development very well. We call it the theory theory. (The theory is that children have theories of the world.)

We think there are very strong similarities between some particular types of early learning—learning about objects and about the mind, in particular—and scientific theory change. In fact, we think they are not just similar but identical. We don't just think that the baby computers have the same general structure as the adult-scientist computers, in the way that perceptual learning and artistic learning and political learning may all have the same general structure. We think that children and scientists actually use some of the same machinery. Scientists are big children. Scientists are such successful learners because they use cognitive abilities that evolution designed for the use of children.

Science also doesn't fit either the caterpillar growth picture or the dinner-bell association picture. It seems extremely unlikely that Einstein's theory of relativity is innately coded in our genes and just happened to mature in Einstein's brain in 1905. On the other hand we've known for a long time that scientists don't just observe the world and write down what they see.

Instead, scientists, like babies, have rich, complex, abstract, coherent representations of the world. They have theories. The theories translate the input—the evidence scientists gather—into a more abstract representation of reality.

Just as children ignore or reinterpret the facts that don't fit their representations, scientists, at least initially, often ignore or reinterpret facts that don't fit their theories. Nor is this necessarily a bad thing. We wouldn't want to rewrite the laws of physics every time an undergraduate screws up in his lab section and gets a weird result. In fact, one advantage of having a theory, for scientists as well as children, is that it lets you know what you should pay attention to.

The theories also go beyond the evidence they are based on. That means they allow scientists to make new predictions about things they've never seen before, just as the children's representations allow them to make new predictions. Those predictions allow scientists, and children, to act on the world in more effective ways.

Just as babies play with the world, testing out their hypotheses on the objects around them, scientists perform experiments. Of course, the scientists' toys are a lot more expensive. All babies need to find out about objects is a set of mixing bowls; to find out about neutrinos you need, quite literally, an act of Congress.

Just as children eventually revise and even replace what they know in the light of what they find out, scientists eventually abandon even cherished theories for new ones. It is true that scientists are less willing to give up their theories than children are, but this may, of course, have something to do with the cost of their toys.

The two most successful examples of human learning turn out to be quite similar. Children and scientists are the best learners in the world, and they both seem to operate in very similar, even identical ways, ways that are unlike even our best computers. They never start from scratch; instead, they modify and change what they already know to gain new knowledge. But they are also never permanently dogmatic—the things they know (or think they know) are always open to further revision.

While the idea that scientists are like children might seem surprising at first, it helps make sense of some otherwise puzzling facts. Scientists, after all, have the same brains as the rest of us. And science is convincing because, at some level, all of us can recognize the value of explaining what goes on around us and predicting what will happen in the future. Yet we have been using our brains to do organized science for only the last five hundred years or so. Why would we have such powerful learning abilities if we never even used them back in the Pleistocene? Where did they come from? And why do there turn out to be such strong similarities between scientists, those great and powerful wizards, and the small and meek little Dorothys we study?

Our answer is that these abilities evolved for the use of babies and young children. In the first chapter of this book we described the evolutionary correlation between prolonged immaturity and cognitive flexibility. We mentioned the idea that, as a species, we trade off the costs of immaturity for the benefits of learning. We human beings are the most cognitively flexible of species, able to cope with the widest variety of environments. In all the debates about whether there was one dispersion of early humans from Africa or many, no one disputes that we did disperse, while our closest primate relatives stayed put. We are infinitely susceptible to the call of the train whistle in the distance. For better or worse we have covered the globe and even made it to outer space. We survive because we change our behavior to suit the particular world in which we find ourselves. We can discover how an ice floe or a desert or South Central L.A. works, and we can change what we do to suit each of these harsh environments.

We also have the most immature and dependent offspring. Parents with college graduates still living in the spare room may occasionally envy the mother cats and father birds who ruthlessly throw their young out after a couple of months. But we know that we couldn't summon up a similar ruthlessness, nor would our babies survive if we did (of course, the college graduates may be a different story). No creature spends more time dependent on others for its very existence than a human baby, and no creature takes on the burden of that dependence so long and so readily as a human adult.

These features of our evolutionary design are consistent with the idea that human beings have unusually powerful and flexible learning abilities. We deploy those abilities during that protected and protracted Eden we call childhood. During our immaturity we don't have to commit ourselves to act in any particular way in order to survive; grown-ups take care of us. That leaves us free to explore many possibilities and to learn just what to do in our particular world. Childhood is a time when we can safely devote ourselves to learning about our specific physical and social environment. We can do pure, basic research while the grown-ups provide the funding and the technology.

For most grown-ups, for most of history, that learning may have largely stopped when we reached maturity and turned to the more central evolutionary business of the four *f*'s (feeding, fleeing, fighting, and engaging in sexual reproduction). We learned most of what we need to know a long time before kindergarten. As adults we can survive in our particular world because as children we figured out how it works.

All the same, the continued existence of these learning abilities allows some of us, some of the time, to continue to learn new things about the world around us. When we give grown-ups leisure and money and interesting problems to solve, they can be almost as smart as babies. We think that, throughout history, some adults continued to learn new things about the world, especially when they were relevant to particular problems of survival. This might explain, for example, the achievements of hunter-gatherer "folk botany" or of Australian aboriginal geography. But the contingencies of history some five hundred years ago gave many more adults the chance to learn about the world. We invented institutions that re-created the conditions of childhood—protected leisure and the right toys. We call those institutions science.

Five hundred years ago a natural activity of children was transformed into an institutionally organized activity of adults. Of course, this transformation led to many differences between what children do and what scientists do. Perhaps the most important difference is that children typically make up theories about close, middle-sized, common objects, including people. As a result they are positively immersed in evidence that is relevant to their theories. Everything they need to know is easily available to them. Scientists, in contrast, often make up theories about objects that are very small or very big, hidden or rare or far away, and the relevant evidence is often very thin on the ground. They make up theories about things such as distant stars and elusive diseases. This relatively small difference has big cognitive and social consequences.

Young children all seem to create similar theories at about the same age. Some developmental psychologists think that this is evidence for the caterpillar growth view. But it is also just what you would expect if children had the same initial theories, had the same mechanisms for revising those theories, and had lots of very similar evidence. Babies around the world start out with the same ideas about people and objects, and they will have similar experiences of people and objects. In every culture different people will sometimes have different beliefs and desires, and objects will continue to exist after they are hidden. These everyday events provide the evidence that lets children revise their initial theories.

In contrast, different scientists, and people in different historical periods and cultures, may have very different kinds of evidence about things like stars and diseases. Often they may not have much relevant evidence at all. They may draw different conclusions as a result. In fact, when scientists start out with the same theories, try to solve the same problems, and get the same evidence—when they're in the same position as babies—they also come up with similar new theories at about the same time. That's why there are all those shared Nobel prizes.

Much of the institutional structure of science is devoted to organizing the search for evidence and evaluating the quality of that evidence. Science demands a complicated division of labor because it takes a great deal of work just to find the right data. Babies and young children do take advantage of other people to help solve problems, but the basic evidence they need is ubiquitous. In contrast, it may take a whole lab full of postdocs and graduate students and research assistants months to find the evidence that is relevant to some scientific hypothesis.

The division of labor also requires elaborate and sometimes fragile social mechanisms for maintaining trust and confidence. Children assume they can trust what their mom tells them. Scientists have to rely not only on their postdocs and research assistants but on competing scientists in other laboratories.

Around 1500 a complex industrial society began to develop,

with important advances in communication and technology. We think that's what made this kind of scientific division of labor possible. We could afford to excuse some people from the requirements of everyday life and let them devote themselves to finding out about the world. Galileo in Italy could rely on the data Tycho Brahe gathered in Denmark and the mathematics Johannes Kepler formulated in Germany. And all of them could rely on the telescope. All these social and technological changes helped the new theory of the planets emerge.

Scientists also seem to invent some new cognitive procedures to deal with problems that nature did not equip us to face. For babies and young children the main problem isn't evaluating evidence, it's making sense of it. Children are mostly concerned about explaining evidence, not deciding if it's reliable. But it's often quite difficult and important for adult scientists to determine whether evidence is reliable. We have to invent detailed experimental protocols and designs to ensure that experimenters in different places will come up with the same results and statistics to allow us to deal with probabilistic evidence. Scientists have to decide when to accept a particular piece of evidence and when to be skeptical about it. Children can afford to be more generally credulous.

These differences between children and scientists are very real. Nevertheless, we think that babies and scientists share the same basic cognitive machinery. They have similar programs, and they reprogram themselves in the same way. They formulate theories, make and test predictions, seek explanations, do experiments, and revise what they know in the light of new evidence. These abilities are at the core of the success of science. All the social institutions would be useless if individual scientists couldn't create theories and test them.

Of course, that means the rest of us have these learning

abilities, too, even if we may not use them as much. We think that these basic abilities are part of our human evolutionary endowment. They allow us to get to the truth about the world because they were designed by evolution to do so. Our eyes are ingeniously and elaborately designed to let us find out about the world. So are our minds.

Explanation as Orgasm

Yet another difference between the biological computers and the man-made computers is that people have emotions and motivations, in a way that man-made computers don't. As *Star Trek* archaeologists, though, we can also investigate the emotions and motivations of the biological computers on this planet. We can even formulate some hypotheses about how those emotions and motivations are related to computational abilities.

We think that babies may even have some of the same emotions and motivations that scientists have. Being a baby may feel like being a scientist. It isn't just that babies can explore and explain their world; they seem driven to do so, even at the risk of life and limb and maternal conniption fits. Like other human drives, that explanatory drive comes equipped with certain emotions: a deeply disturbing dissatisfaction when you can't make sense of things and a distinctive joy when you can. We can actually see those emotions on the faces of babies and young children. They purse their lips and wrinkle their brows when we present them with an object-permanence or false-belief problem, and then produce a radiant, even smug smile well before they actually give us the right answer. (We used to recruit many of our baby subjects through the La Leche League, a group that encourages breast-feeding. Often when we presented these babies with a really tough problem, they would frown, turn to their mothers for a quick, comforting snack, and then, refreshed and cheerful, turn back to the problem—one explanation for the cigarettes and candy bars of the late-night lab.)

Even busy grown-ups, preoccupied with the modern versions of the four *f*'s (increasing your salary, intimidating your bureaucratic enemies, outwitting the ones you can't intimidate, and flirting over the watercooler), know the satisfaction of figuring things out. Our pleasure in reading popular science is a kind of vicarious pleasure in seeing how problems can be solved and oddities can be explained. In our spare time we set ourselves problems, from chess to crosswords, and take pleasure in their solution. When we get the chance, we play, too.

While professional scientists may be driven partly by greed, ambition, anxiety, lust, and other grown-up drives (the four *f*'s go on in every laboratory), the explanatory drive also plays an important role. The physicist Steven Weinberg put it this way: "Nature seems to act on us as a teaching machine. When a scientist reaches a new understanding of nature he or she experiences an intense pleasure. These experiences over long periods have taught us how to judge what sort of scientific theory will provide the pleasure of understanding nature." The NASA scientists conducting the Mars probes expressed their delight by saying that they felt as if they were little kids again. None of them said they felt as if they just got a raise.

Explanation is to cognition as orgasm is to reproduction: it is an intensely pleasurable experience that marks the successful completion of a natural drive. As the seventeenth-century philosopher Thomas Hobbes put it: "There is a lust of the mind that, by a perseverance of delight in the continual and indefatigable generation of knowledge, far exceedeth the short vehemence of carnal pleasure" (perhaps a slight exaggeration). We think that these distinctively human cognitive emotions—the agony of confusion and the ecstasy of explanation—may be the mark of the operation of the natural cognitive system that allows us to learn when we are very young.

It may seem to us that we make up theories of the world because we want explanations, just as it seems to us that we have sex because we want orgasms. From the evolutionary point of view, though, the relationship is the reverse. Orgasms guarantee that we will keep trying to have sex, and our joy in explanation guarantees that we will keep trying to construct better, truer theories of the world. Getting the world right, like having sex, gives us a long-term evolutionary advantage. Drives and emotions turn those long-term advantages into short-term motivations. All of us are driven by these cognitive emotions sometimes, scientists are driven by them much of the time, and babies, who have so much to learn, are in their grip practically all the time.

Studying babies makes us realize that the biological computers on this planet differ from the man-made computers in this regard, as well. They don't just compute, learn, reason, and know. They are driven to do all these things and are designed to take intense pleasure in doing so.

Other People

Yet another difference between the man-made computers and the biological computers is that the biological computers directly influence one another's programs. They are designed to work as part of a complex social network. How does this affect the way we learn? How much do children learn about the world themselves, and how much are they taught?

Parents, most definitely including the parents writing this book, tend to waver between a manic, megalomaniac certainty that everything depends on them and a depressive, crushed sense of their own helplessness. Developmental psychology has wavered in the same way. Some theories discount the influence of other people. Obviously, this is true of the caterpillar theories. If most of what we know is the result of a genetic blueprint, then there isn't much room for parents and other people to have an influence. Piaget's theory also tended to discount the influence of other people. In his anxiety to emphasize the child's role in development, Piaget deemphasized the role of grown-ups. (When Piaget took over the center at Geneva, he changed the school's logo. It had shown an adult leading a child; he changed it to a child leading an adult.)

On the other hand, other traditions, such as behaviorism, tended to make parents and other caretakers central. They were responsible at once for everything right and for everything wrong about their children. Perhaps these views retain their appeal because they absolve adult children of responsibility (they can always blame it on Mom) and also give adult parents a sense of power.

Nurture as Nature

All of these theories and debates presuppose that there is a deep separation between a "natural" biologically determined part of knowledge that comes from genes and a "cultural" socially determined part that comes from parents. The new developmental research undercuts this distinction. The interactions between children and adults seem as natural and deeply ingrained as anything else about us.

Take something as simple as the fact that babies are so incredibly cute. Babies' cuteness turns out to be a deeply ingrained biological fact, a fact that is equally about babies and about us. Some of it is simply physical. The features of babies' faces—the large, bulging foreheads and big eyes and little mouths and chins—automatically call up positive reactions in adults. And they evoke love not just in mothers (who, after all, have been known to love faces that only a mother could love) but in almost everyone. You can invent what biologists call a supernormal baby stimulus, an unreal face in which all these features are exaggerated, and people will react even more positively; they'll think it's supercute.

Hollywood unconsciously exploits this fact. The extraterrestrial E.T., in spite of his superficially weird and alien appearance, is actually a supernormal stimulus, with exaggerated versions of baby features. So are the Ewoks in the *Star Wars* movies. Part of our evolution is this coordination between what babies look like and what grown-ups think is cute. That coordination leads grown-ups to provide an environment in which babies can flourish.

The cuteness effect, though, isn't all physical. Consider imitation. We saw that even very young babies can imitate grown-up expressions. In *E.T.* both Elliott, the boy hero, and the viewers are drawn to the alien as soon as we see his cute face. But the real link comes when E.T. begins to imitate Elliott's actions. That interaction tells Elliott and us that E.T. is a creature with a mind, a creature that is somehow like us.

Seeing a very young baby imitate you has a similar effect. Suddenly you are in tune, hooked up, *en rapport.* This small, strange being starts to make sense. But imitation is not a game you can play by yourself. It takes two.

Babies themselves seem to like mutual imitation games as much as adults do. Not only do babies imitate adults, adults quite unconsciously imitate babies. Mothers open their own mouth as they place a spoon in their baby's mouth. Moreover, Andy discovered that babies know when adults are imitating them and they like it. He was inspired by the old Marx Brothers routine in which Chico stands on one side of a mirrorlike doorway and imitates all of Groucho's actions. Andy did an experiment in which an adult was "yoked" to a baby like a human mirror. The adult did everything the baby did. When the baby banged the table, the grown-up did, too; when the baby raised his arm, so did the adult. Another adult was told to do something different from the baby each time: when the baby raised his hand, the adult banged the table and so on. One-year-olds consistently preferred to look at the adult who mimicked them. Moreover, they produced "testing" behaviors, just the way Groucho did in the routine. They would produce some weird action just to see what the grown-up would do.

As in the case of simple physical cuteness, both grown-ups and babies seem to be biologically designed to engage in mutual imitation. But there is also an interesting difference between these two phenomena. The biological preference for baby faces seems to be instilled by nature; it's a kind of instinct, and, of course, the babies' faces themselves are determined by their genes. But imitation actually leads babies to behave in new ways that are *not* genetically determined and, in fact, to behave like the adults around them. Imitation is the motor for culture. By imitating what the particular adults around them do, young children learn how to behave in the particular social world—the particular family or community or culture—they find themselves in. They can draw a bow or dress a doll or even learn such bizarre cultural rituals as pulling a piece of toothed plastic through their hair every morning and rubbing a stiff brush against their teeth every night.

Babies and young children can also use imitation to learn important new things about how the physical world works. In Chapter Two we described how one-year-olds who see a grown-up press his forehead to a box to make it light up will do the same thing themselves. Human babies who see an adult use a tool in a particular way will learn how to use that tool themselves. And by imitating the babies' actions, often with some grown-up variation, the grown-ups can show the babies what they ought to be doing. Imitation is an innate mechanism for learning from adults, a culture instinct. In fact, recent research suggests that most other animals don't learn through imitation in this way.

Motherese is another good example of the way adults are designed to help babies learn. Adults unconsciously produce this special type of language when they are talking to babies. Motherese captures the babies' attention; babies seem designed to like to listen to this kind of speech. But motherese also makes the sound structure of the language particularly clear. Recall that Swedish, English, and Russian mothers each produced different types of vowels, vowels that were helpful in pointing to the structure of their own particular language. Babies seem designed unconsciously to use this information to crack the speech code. Just as imitation helps children learn the specific ways people act in their culture, motherese apparently helps babies learn the specific sounds of their own language.

Motherese seems to teach children about people and objects as well as about words. Korean-speaking mothers emphasize action when they speak to their babies, while English speakers emphasize objects. Just as babies seem responsive to the differences in sounds, they also seem responsive to these differences in content. The Korean-speaking babies seem to focus more on actions, while the English-speaking babies pay more attention to objects. The grown-ups' language seems to lead the children toward new ideas about the world.

One important aspect of the new research is that these social influences come from other people in general, not just from mothers. In practice, of course, mothers actually do most of the work of raising children, and that is why they've been the focus of research. (It is also, apart from euphony, why we use the word *motherese* rather than *caregiverese* or *child-directed speech.* God knows, mothers don't get any money or fame or power for the work they do; at least they ought to get a little terminological recognition from psychologists.) But none of the phenomena we've discovered seem to be exclusive to mothers, or even really to grown-ups. Anyone who talks to a baby is likely to play imitation games and to use motherese. Andy found that babies would imitate other children as well as adults. Even four-year-old children use a kind of motherese when they talk to their baby brothers and sisters. In fact, we saw that sometimes older siblings seem to be even more important than parents. Children with older siblings seem to learn about the differences between their own mind and the minds of others more quickly than only children.

The second important thing about the influence of other people is that the most significant behavior seems almost entirely unintentional. Parents don't deliberately set out to imitate their babies or to speak motherese; it's just what comes naturally. Our instinctive behaviors toward babies and babies' instinctive behaviors toward us combine to enable the babies to learn as much as they do.

The third important thing about the influence of other people is that it seems to work in concert with children's own learning abilities. Newborns will imitate facial expressions, but only much older babies will imitate actions on objects, like touching their forehead to the box. Babies won't imitate complex actions they don't understand themselves. At first, young three-year-olds stick to their claim that they said there were pencils in the box even when you tell them they're wrong. Children won't take in what you tell them until it makes sense to them. Other people don't simply shape what children do;

parents aren't the programmers. Instead, they seem designed to provide just the right sort of information at just the right time to help the children reprogram themselves.

The Klingons and the Vulcans

These questions about children's knowledge reflect broader questions about knowledge in general. Just as there are debates about how much knowledge we are born with and how much we learn, there are debates about how much individuals and society contribute to the development of knowledge, particularly scientific knowledge.

Originally, many philosophers tried to show how an ideal scientist should think about the world. They thought of science as a detached, idealized enterprise that used logical procedures to get to the truth. But when you actually looked at how real scientists did real science, this empyreal ideal seemed to vanish in a welter of academic politics, intellectual alliances and enmities, and naked ambition. Historians and sociologists of science pointed to the complex networks of power and influence that were involved in scientific change. Real science looked more like the brutal intrigues of the barbarous *Star Trek* Klingons than like logical Mr. Spock among the Vulcans.

The reaction produced "postmodernist" theories of science. According to these theories, science doesn't get to the truth at all. The truth is simply what a bunch of the most powerful scientists decide it is. The assumption seems to be that if scientific change involves social interaction, then, for that very reason, science can't approach the truth. But this nihilist answer not only flies in the face of intuition, it also leaves us with a terrific puzzle. According to the postmodernists, science doesn't tell us about the real world at all. Instead, it's just a social agreement, an elaborate comedy—or perhaps a Klingon tragedy—of manners. But then why do scientific explanations convince the rest of us? And even more dramatic, how is it that we can use scientific theories to get real rockets to a real moon?

If scientists are like children, then they are neither the chilly, idealized, logical Vulcans of philosophical legend nor the brutal Klingons in lab coats of postmodernist myth. The sociologists were right that science is a social enterprise—we are an intrinsically social species, and all our successful projects depend on cooperation with others. But the philosophers were right that science is a logical enterprise—we are also a species that intrinsically seeks the truth, and we have powerful reasoning abilities that let us find it.

These two aspects of our nature may be in conflict some of the time, but most of the time, and especially in childhood, they are in concord. Babies depend on other people for much of their information about the world. But that dependence makes babies more in tune with the real world around them, not less. It gives children more and better information about the world than they could find for themselves. As children we depend on other people to pass on the information that hundreds of previous generations have accumulated. Together the children and the grown-ups (and other children) who take care of them form a kind of system for getting to the truth.

At its best science works the same way. Scientists, like children, depend both on their own individual theory-formation abilities and on a social network of shared information. But, contrary to the postmodernist view, this makes science more likely to get to the truth, not less.

The same is true for other adult endeavors. The debates about how babies and scientists learn are similar to debates about art and politics. There are comparable arguments about whether there are universal, inalterable artistic and political principles or whether artistic and political values are simply relative to a particular culture. Participating in an artistic or political community and tradition is obviously important; the lone artistic genius in the garret or the lone heroic leader is as much of a myth as the lone scientific genius in the lab. But that also doesn't mean that artistic or political value is just whatever powerful artists or politicians say it is.

Sailing Together

Let's go back to the original question about how children solve the ancient problems of knowledge so well and so quickly. Our first pass at an answer to that question is to think of children as if they were special biological computers. That helps us understand babies and young children in a new way and appreciate the sophistication of their intellectual abilities. Even the youngest babies are already born with powerful programs for interpreting the world.

But that also helps us think about computers in new ways. The babies are unlike any computers we know. The babies can change their own programs. They have emotions and drives that actively cause them to explore the world and learn more. And they get much of their information from other people who are, in fact, designed to fulfill just this purpose.

If babies have these abilities, so do we. We see these abilities at work when we adults get a chance to engage in the kind of learning that is so characteristic of babies and young children, particularly when we do science.

Studying babies and young children, then, gives us a new view of how we adults come to understand the world and of how a new type of computer might be designed that could understand the world as well as we do. The details of that view are still sketchy, there is still much work to be done, but the broad outlines are beginning to emerge. We begin with representations of the world, and in concert with the other people around us, we alter and revise those representations. This process can go on indefinitely. We developmental scientists and the children we study, we parents and the children we love, sail in Ulysses' boat together. While it is often a wild, frustrating, and difficult journey, and we are rather likely to feel we are in over our heads, at least we couldn't ask for better company.

CHAPTER SIX

What Scientists Have Learned About Children's Brains

All the things we've described so far—our minds, our knowledge, our perceptions, predictions, thoughts, and emotions—ultimately depend on three pounds of quivering gray jelly. The gray jelly has been the center of a lot of attention lately. Brains are suddenly sexy (which just goes to show that sometimes neither size nor looks matter all that much). Magazine stories describe "critical periods" for learning and "cell death and synaptic pruning" in the infant brain. When Pat gives talks about language, she invariably fields questions such as "If I don't give my child the right language experiences now, will it be impossible for his brain to take it in later?" Or "Is there anything I can do to stop my baby from losing brain cells?"

We actually know much more about how the mind develops than we do about the brain. There is a tendency to think that changes in the brain must somehow cause changes in our knowledge, that there is, say, some physical change that causes babies to understand things in a new way at eighteen months. But it would be just as accurate to put this the other way around. Babies' brains change as a result of the new things they learn about the world. Studying babies' minds, in the ways we've described so far, *is* studying babies' brains and is, so far, the most productive method of studying their brains we know about. Understanding the mind helps us understand the brain as much as or more than understanding the brain helps us understand the mind. And in fact, neuroscientists have begun to make progress in understanding the brain by linking the physiology to what we know about the mind.

There is also a tendency to think that if something is encoded in the brain, it must be genetically determined. But again if you think about it for a minute, you can see that that can't be; everything that happens to us must make a difference to our brain if it's going to make a difference to our mind. Babies certainly don't learn through changes in their big toes. As neuroscience techniques have improved, they have begun to show how, and how much, our brains change, and how much that change is due to our experience of the world. We still don't know very much about how the programs we described in the last chapter are physically coded in our brains. But we are starting to understand more about how the adult brain works and how much the baby's brain changes.

The Adult Brain

Pat shudders when she remembers her days as a graduate student in the neurology ward at a VA hospital. Thursday mornings were particularly difficult. She didn't eat breakfast on Thursdays because at 7:00 A.M. the Neurology grand rounds began, and she would spend them looking at slices of brain tissue donated by someone who had recently died. Many of these brains came from patients she had studied who had the language disorder known as aphasia. The goal was to find out where the brain had been damaged.

Pat particularly remembers a brilliant pediatric neurosurgeon who suffered a very severe stroke. His vivid intelligence was still there, and he could communicate a bit with his facial expressions and gestures, but he couldn't speak and understand or read or write. His language had disappeared. All he could say was "ta, ta, ta." One Thursday morning his brain, neatly sliced in millimeter sections, was part of the grand rounds.

This anatomical work suggested that the adult brain was made up of highly specialized parts. When a particular part of the brain was damaged, the patient lost a particular mental ability. In right-handed people the left side of the brain is specialized for language, the right side perceives faces and music, and the back of the brain is responsible for vision. The mute neurosurgeon's brain had been damaged in one particular part of the left hemisphere. The intact parts of his brain allowed him to think and feel and remember normally, but the damage made him lose the ability to speak.

All of this work was based on studies of dead brains. Dead brains don't look very dynamic. Today, new techniques let us look at living brains. These techniques reveal that the brain is specialized in a different way. The anatomical studies of damaged brains showed that certain areas of the brain are necessary in order to speak or to see or to recognize people. But it turns out that even individual living cells are highly tuned to specific kinds of information.

In these studies, scientists "listen" to a brain at work. The scientists insert a tiny electrode (which isn't painful) into an individual cell in an animal's brain. The electrode records the activity of the cell as the animal looks at objects or listens to sounds. When a cell "likes" a particular stimulus, say a picture the animal is looking at, the cell starts creating electrical impulses that sound like static on a radio. The greater the electrical activity, the greater the noise. Scientists sit for hours listening to the cells tell them what they like to see and hear.

Cells are extremely picky about what they respond to. Certain cells respond only to faces, sometimes only to particular faces, or to faces in general that are oriented in a particular direction (sideways, for example). Other cells respond only to visual movement in a certain direction or a shape of a particular kind. Some cells in the auditory cortex respond to notes of a particular frequency, and the tuning is quite precise. Others respond to sounds of a particular loudness or sounds that change in frequency, swooping up in pitch or down. There even seem to be cells that respond both when an animal makes a particular movement and when it sees another animal move in the same way. When a cell hears or sees the kind of stimulation it is tuned to, it creates a burst of electrical activity that communicates what is out there to other parts of the brain.

Groups of individual cells firing in this way can work like the circuits in a computer. You could think of these cells as devices that translate the information at our senses into a more abstract category or symbol. The cells on the retina respond to light, and they send a visual pattern to the brain, like a video image. But then other cells higher up in the brain respond to some of these patterns and not others. These cells seem to say "This is a face" or "This is a moving object." Then they send that more abstract information to other parts of the brain. Complicated arrays of cells like these could allow babies to recognize that faces are special or that moving objects are different from still objects or even that their own facial expressions are like the facial expressions they see other people make.

We can use other methods to look at living human brains. Neuroscientists have developed new techniques that effectively let us look at the brain as it functions. These techniques produce the spectacular photographs that show brains "lit up" in different colors while their owners think. They permit us to study the brains of living, conscious people, people who can report the mental gymnastics they are doing while we watch their brains doing them.

These new techniques allow scientists to look globally at large areas of the brain, rather than at a single cell, and see how the brain works in concert. They show which brain areas are active when a person remembers the tune of "Yesterday," hears a sentence about the World Series, sees a reproduction of a Monet water lily, or thinks about his mother. Either these methods record the brain's electrical activity through the scalp, or they record the brain's metabolic activity as it burns glucose, the body's fuel.

You can track the brain's electrical activity from outside the skull with electroencephalograms (EEG), event-related potentials (ERP), and magnetoencephalography (MEG). In these experiments the subject wears a huge cap containing anywhere from twenty to two hundred electrodes (sometimes referred to as the hair dryer from hell) to pick up brain waves. As the person underneath the cap thinks about a certain thing or looks at a particular picture, the cap records the electrical activity of the person's brain. Pat even tests young babies this way. Their caps are little and soft, like baby bonnets, and contain only twenty electrodes, but that is enough to record activity in the babies' brains from many locations. Pat tests babies who listen to speech sounds through an earphone. The babies are completely unaware that we are recording their brain waves while they listen.

Other techniques, such as PET (positron emission tomography) and fMRI (functional magnetic resonance imaging), measure brain activity even more directly. PET exploits the fact that the more active parts of the brain burn more glucose, the brain's fuel, than the less active parts, just as a more active muscle will burn up more calories than a less active one. In PET scans the brain is injected with radioactive glucose, and the scan traces which parts of the brain are burning up this glucose. The technique has been used primarily with patients who require brain surgery; it helps locate important structures that need to be avoided during the operation. The fMRI technique also measures which regions of the brain are most active, and does this by tracking blood flow and oxygenation. The fMRI doesn't involve injections and can even be used with children.

Both of these techniques show activity in the brain more directly and in more detail than the methods that record from outside the brain. As the subjects do different things, listen to sounds or look at pictures or even just think, we can see which parts of the brain are getting a workout. The scans show that different brain areas are active when we think about different things. As we might expect from the studies of damaged dead brains, language activates the left hemisphere, faces activate the right hemisphere, and visual patterns activate the back of the brain.

The new studies also show the intricate interplay between different areas of the brain and how it unfolds over time as we see, hear, and think. The different areas are coordinated when we have any particular experience. When you see a cow mooing, your visual areas light up as they record its black and white color, your auditory and language areas light up as you hear it moo and think, "It's a cow," and your taste and smell areas light up as you think about milk and the smell of a cow and a farm. In a split second the brain takes the stimulus apart and puts it back together again.

Two things emerge from all these studies. The adult brain is a highly specialized device that responds specifically to specific kinds of stimulation. Particular parts of the brain, even individual cells, are designed to respond to information from the outside world in particular ways, sending that information off to other parts of the brain. In that sense the brain is like a classical computer. The brain is also, however, a dynamic and active system. Its parts are constantly interacting with one another, and often many parts of the brain and certainly many, many cells are simultaneously involved in processing even a simple piece of information. Unlike most computers, the brain has no single place where all the decisions are made or where all the information is stored.

How Brains Get Built

Where does all this specialized structure come from? It could, of course, all be built in to begin with. That would be the neurological equivalent of the caterpillar genetic-blueprint theory of the mind. Some structure clearly is built in, but the brain also seems to change radically in response to experience. This has become increasingly apparent as our techniques for studying the brain have improved. We used to think the brain's physical structure developed on a fixed schedule, more or less independent of what happened in the baby's world. We know now that there isn't a genetic blueprint that simply ticks off the milestones of brain development one by one.

The electronic computers we have now don't work until after they are built. They are assembled using chips and circuits according to a very specific diagram. When all the connections are soldered together, we turn on the computer for the first time and it begins to work (or at least so we hope). The hardware of the computer doesn't change as we use it any more than the wiring on a Christmas tree changes when we flip the switch and the lights go on.

But the human brain works very differently. The brain keeps rewiring itself even after it is turned on. And the circuits that are laid down depend deeply on experience. Experience is changing the brain from the very beginning. Everything a baby sees, hears, tastes, touches, and smells influences the way the brain gets hooked up.

If the computer on your desk worked like a brain, it would get better and better at word-processing your manuscripts the longer you let it run and the more words you typed into it. After you had used it for a few years, all you might have to do is type in "Write *Scientist in the Crib,*" and it would do the rest. (We wish.) If you opened it up after those few years, the silicon chips and circuits would be arranged in a completely different way than they were when you bought it.

Because we actually participate in building our own brains, and because each of us has a unique history of experiences, each brain is unique. Eventually the adult brain becomes a complex thicket of particular connections. Estimates are that it takes a quadrillion connections—that's 1,000 trillion—to wire an adult brain. Your specific pattern of connections, the wiring diagram of your brain, defines you as an individual person. It's as if we each have our very own custom-designed utterly unique program, Pat Kuhl 99 or Alison Gopnik 43.5. In Alzheimer's disease, this thicket of connections gets undone, the brain's mental capacities fade away, and the individual person begins, tragically, to fade away, as well.

The most dramatic evidence that the brain is like this comes from animal studies. In the first studies, thirty years ago, neuroscientists discovered that laboratory rats raised in a "rich" environment, with wheels to spin, ladders to climb on, and other rats to play with, grew thicker brains than rats who were raised alone in a lab cage with no playmates or toys. After the rats spent two weeks in the cage with toys and playmates, the brain areas involved in sensory perception were 16 percent thicker.

The rats with thicker brains were smarter, too. They figured out how to run mazes and find food faster than those with thinner brains. The brains of the deprived rats got smaller. There was even an effect on the next generation: pregnant rats who lived in the rich environment produced newborn rat pups with a thicker cortex than those who lived in the impoverished one.

It's important to point out that in this experiment the "rich" environment in the laboratory was actually more like the normal environment of a rat in the wild. A rat living in the sewers of New York City survives by working out where the best garbage is, fighting and mating with other rats, and cleverly eluding exterminators. It may not be a great life, but it's certainly stimulating. So rather than saying the studies showed that extra stimulation led to a thicker rat brain, it might be more accurate to say the studies showed that a more normal environment leads to a thicker brain than a deprived environment.

The same thing seems to be true of human babies. The new scientific research doesn't say that parents should provide special "enriching" experiences to children over and above what they experience in everyday life. It does suggest, though, that a radically deprived environment could cause damage.

Other experiments showed that the effects of experience could be much more precise and specific. In one classic experiment, the Nobel Prize winners David Hubel and Torsten Wiesel covered one eye of newborn kittens. The kittens continued to do all the normal things that kittens do, but with one eye rather than two. After several months, the scientists uncovered the eye and looked at the connections between the two eyes and the brain. The surprise was that if the eye had been covered beyond a particular amount of time, it was effectively blind. It was not connected to the brain. This despite the fact that the eye was perfectly normal from an optical point of view.

What had happened was that the brain had gotten no stimulation from that eye, and it had wired itself to receive information only from the other, open eye. There had been a sort of takeover: all the brain cells in that portion of the brain had been taken over by one eye, leaving no connections to the other one.

This early research with animals established an important point—a brain can physically expand and contract and change depending on experience. Modern neuroscience has gone a long way toward explaining why that's true.

Wiring the Brain: Talk to Me

An adult brain has about 100 billion nerve cells, or neurons, about the same number as the number of stars in the Milky Way. A baby's brain contains most of the neurons it will ever have. The number of neurons remains very nearly the same from the time we're born until we're well past sixty-five years old. But the newborn brain weighs only about one quarter as much as the adult brain. What grows, what changes?

Neurons grow, and that accounts for some of the difference, but mostly what changes is the wiring, the intricate network of connections between cells. Those connections are what allow an individual cell to respond in particular ways to other cells. The connections, for example, allow an individual cell in the cortex to respond when, and only when, the cells of the retina send it a picture of a face.

This intricate wiring depends on activity and experience. Think of it this way. Cells grow in different parts of the brain. In order to influence one another, these cells have to talk to one another. Communication is difficult because the distances are large (in neuronal terms). When we want to talk to people far away, what do we do? We have to make some long-distance connection—like a telephone call. Neurons do the equivalent of growing telephone wires that allow them to communicate with and influence one another. Rather than waiting, like the passive computer, for a technician to hook them up, they physically grow their own connections to other cells.

New techniques allow neuroscientists to closely examine the brain's cells during their early embryonic development, as the brain begins its primitive wiring. This research shows that the wiring is "activity dependent": brain cells get connected by sending out electrical signals. Even before birth, brain cells are spontaneously firing, sending off bursts of electricity and trying to signal one another. Scientists compare it to autodialing a telephone. Groups of cells send signals in waves trying to reach other cells. Cells that fire at the same time grow connections to one another (a favorite phrase of neuroscientists is "Cells that fire together wire together"). Evidently, even cells want to be in touch with others who respond to them.

After birth, as experience floods in from all the sensory organs, cells keep trying to make connections with one another. It's no simple feat. The nerve cells of the eye, for example, have to connect to the optic nerve and eventually to the visual centers all the way at the back of the brain. And those cells have to bypass the brain centers responsible for hearing and touch in order to reach their destination. It's like stringing telephone lines between particular homes in specific cities.

This pattern of growing and connecting cells isn't completely random, but it's also far from predetermined. Studies of animals show that some instructions are laid down by the genes, like the basic telephone trunk lines laid between cities. Cells in the retina of the eye do send connections out toward the visual areas in the back of the brain, rather than to the language centers on the sides of the brain. But, beyond that, the wiring depends on activity. The basic trunk lines are laid down, but the specific connections from one house to another house require something more.

As cells signal to one another, they lay down these more permanent connections. It's as if, when you used your cell phone to call your neighbor often enough, a cable spontaneously grew between your houses. At first, cells exuberantly attempt to connect to as many other cells as they can. Like phone solicitors, they call everyone, hoping someone will answer and say yes. When another cell does answer, and answers often enough, a more permanent link gets laid down.

Making these permanent connections is what brain cells live for. As a cell matures, it sends out multiple branches trying to make contact with other cells. Some branches (called axons) send information out of the cell, and others (called dendrites) send information into the cell. The object is to connect one cell's axon to another cell's dendrite. The connection between the two is called a synapse. When an axon reaches a dendrite, it sets up a special kind of communication—neurotransmission. When two neurons form a synapse, chemicals can flow between them and the connection is complete: the calls can go through.

So synapses in the brain are the long-term connections that

allow cells to talk to one another. That means that the number of synapses gives us a rough estimate of the baby's progress in wiring the brain. Remember that nerve cells in action burn up fuel, just like muscles. If we measure the amount of glucose metabolism in the brain, we can estimate how many synapses are operating at different times during development.

The news is that children's brains are much busier than ours. By three months the brain areas involved in seeing, hearing, and touch are burning up an increasing amount of glucose. The brain's energy consumption reaches full adult levels at around two years of age. By three the little child's brain is actually twice as active as an adult brain. This bristling activity remains at twice the level of an adult until the child reaches the age of nine or ten. It begins to decline around then but reaches adult levels only at about eighteen.

What's behind all this furious activity? The brain is busy setting up its connections. At birth, each neuron in the cerebral cortex has around 2,500 synapses. The number of synapses reaches its peak at two to three years of age, when there are about 15,000 synapses per neuron. This is actually many more than in an adult brain. Preschool children have brains that are literally more active, more connected, and much more flexible than ours. From the point of view of neurology, they really are alien geniuses.

Synaptic Pruning: When a Loss Is a Gain

What happens to all these connections as we get older? Brains don't just steadily make more and more connections. Instead, they grow many more connections than they need and then get rid of lots of them. It turns out that deleting old connections is just as important as adding new ones. The synapses that carry the most messages get stronger and survive, while weaker synaptic connections are cut out.

This process is rather like pruning a fruit tree or pinching a geranium plant. Stopping the growth in some branches strengthens the growth in other branches and changes the whole design of the plant. The brain can make a frequently used connection stronger by pruning connections that aren't used. Experience determines which connections will be strengthened and which will be pruned; connections that have been activated most frequently get preserved. Between about age ten and puberty, the brain will ruthlessly destroy its weakest connections, preserving only those that experience has shown to be useful.

The "loss" of brain connections that Pat's audience worried about is actually a very good thing. It allows the highly specialized adult brain to be finely attuned to its particular environment. The brain is very flexible. It has what neuroscientists call plasticity. The process of connecting and pruning allows the brain to adapt itself to its surroundings. That's how our ancestors could survive in the savanna as well as the forest, how we can survive in our modern-day jungle, and how our grandchildren may be able to survive in outer space.

This cycle of physical growth and pruning in the brain may be related to the changes in children's knowledge we described earlier. Remember the change in the discrimination of speech sounds that takes place during the first year of life? At birth the baby's citizen-of-the-world brain recognizes the subtle differences among all sounds of all languages. But in order to acquire a specific language, the infant brain has to develop a structure that emphasizes the distinctions in the child's own language and ignores others.

The exuberant connections of early infancy may allow all those sounds to be discriminated. But then, particular words and sentences in a particular language pour into the baby's brain. As the brain processes all these sounds, the way the baby perceives sounds is reorganized. It's possible that the way this happens is that the brain prunes connections that never get used and creates and strengthens and sharpens connections that are often stimulated. Certainly as we play with our three-month-old babies, cooing back and forth and talking nonsense, and also producing hundreds of thousands of *ahh*'s and *ooo*'s, we are altering the babies' brains.

There is even a bit of experimental evidence to show that changes in the way babies hear sounds are linked to changes in their brains. In an ERP study, infants wore the baby bonnets with electrodes in them. Early in development, the babies' brains showed no difference in reaction to prototypical sounds from their own native language or a foreign language. But just a few months later these same babies showed unique patterns of brain activity in their left hemispheres when they heard prototypes from their own language. The brain physically changed as infants learned their native language.

More abstractly, we can see how the same sorts of processes might also be involved in something like changing a theory. Changing a theory involves both adding new links between ideas and getting rid of earlier ideas that turned out to be wrong.

While we were once ready to learn anything (or at least many things), we become more specialized with age, less ready to learn something new, more set in our ways. But we also become much more skillful and competent—we do the things we can do much more quickly and easily, from speaking to reading to tying our shoelaces to writing books. Buying that same kind of black dress over and over again may look like a lack of imagination, but it may actually be the result of knowing what works—a hard-earned wisdom that comes only after we've given closetfuls of lime-green capri pants and hot-pink tank tops to Goodwill.

The brain seems better than most of us at getting rid of unused clutter. It throws out the things that don't work and keeps the things that do. The process of pruning leaves the adult brain much more highly specialized than the baby's brain, with particular activities more confined to particular areas and compartments. This specialized structure enables us to do all the things we can do as adults.

Even when we are adults, though, the process of making new connections, pruning old ones, and generating new brain cells continues to go on. That process allows us to remember new things and forget old ones, and to learn how to do new things and to develop new ideas. Even in middle age we may occasionally discover that, actually, that new skirt works a lot better than the old one. And we hope that even an activity like reading this book may lead you to delete a few old connections and make some new ones.

Are There Critical Periods?

All of this research is consistent with the idea that childhood is the time when we learn most and when our brains as well as our minds are most open to new experience. We saw that this is also the picture that comes out of psychological studies of development. Babies and young children are perpetually exploring and experimenting, testing out new theories and changing old theories when they learn something new. Although the process doesn't stop in adulthood, it certainly slows down.

But some researchers have also suggested a stronger version of this idea. They suggest that the baby's brain is open to experience of a particular kind only during narrow periods of opportunity. It is as if there is an open door to the brain and experience rushes through it for some special critical period of time, but then suddenly the door slams shut, or as if you can make deposits at the brain bank only until five o'clock, when the teller's window closes. The brain learns only during this critical period.

These questions about brain development and the timing of experience are important for neuroscientists, but they are also questions that have an impact on real life. If children don't hear the right kind of speech before they go to school, will they be able to learn to speak normally? If babies grow up in loveless orphanages, will they be emotionally scarred for life?

Everyone agrees that the neural sculpting that goes on during the early phases of development is unique and deeply influences the rest of development. The question is whether there is a specific "clock" that determines when experience will be useful. How strictly timed are the learning opportunities?

There are, in fact, plenty of examples from animals in which experience must indeed be timed quite precisely to have an impact, almost like a biological alarm clock. When that time has passed, the biological clock rings, the door slams shut, and experience is no longer effective.

For example, experimenters tried playing tape-recorded bird songs to birds at various ages. In the white-crowned sparrow the window for learning a song was open for only about thirty days (between twenty and fifty days of age). If the male bird didn't hear the right song in this period, he wouldn't be able to sing normally and wouldn't be able to attract a mate and breed. (Learning to sing is pretty important to a male bird—it's looks, wealth, and power all wrapped up in one set of tweets.) Hearing the song later on didn't help.

Another dramatic example comes from the studies we mentioned earlier, which showed that experience is required to wire the connections between the two eyes and the brain. Hubel and Weisel found that the input to the kittens' eyes had to come at a very specific time, between thirty and eighty days after they were born. If an eye was open during that period, it would be wired correctly to the brain. If it was covered during that period and uncovered only after eighty days, it was too late: the other eye had taken over, and the unused eye would continue to be blind for the rest of the animal's life.

Although these examples of critical periods appear to involve a biological clock, new discoveries indicate that in other cases, something else may also be going on. In these cases, early experience also has a profound and long-lasting effect, but the new view is that they aren't ruled entirely by the clock. It isn't just that maturation brings the period of flexible learning to a close.

Instead, it may be that experience itself has changed our brains so that we perceive and interpret the world in a certain way. Once the neural wiring occurs, it is difficult to interpret the world in any different way. Once we have a representation that works, and instances mount up that confirm that representation, it becomes increasingly difficult to change it. When we are quite sure something is true, we are less likely to be willing, or even able, to change our minds about it, and this also seems to be true of our neural representations. In both the classical and the new view, people can learn more easily at one age than another. The crucial difference is whether this is due to a biological clock or to the brain structures we have already developed.

These two interpretations are being hotly debated in the case of language learning. Some scientists believe there is a critical period for language acquisition in humans—a biological clock that cuts off our ability to learn later on. The most dramatic evidence of this comes from the terrible natural experiments that create "wild children," children who don't hear language early in life.

Thirty years ago social workers discovered a fourteen-year-old California girl named Genie. For most of her life, Genie's father had tied her to a chair in a small room and kept her away from the rest of the world. If she raised her voice above a whisper, she was beaten. Genie seemed unable to acquire normal language even after she had been rescued. But, of course, Genie suffered in all sorts of ways from her inhuman treatment, so it's hard to isolate the precise factors that affected her language.

Other, less tragic kinds of evidence also support this idea. Most people have a much more difficult time learning a second language late in life than they do in childhood. Immigrants may try to learn the language of their new country, only to be outdone by their own children. When we visit a foreign country for a while, our kids seem to be happily chatting with the other kids in the playground, while we are still painfully looking through the phrase books. When we learn a second language past puberty, we speak with a foreign accent—in other words, with phonetics, intonation, and stress patterns that are not appropriate for the new language. We also have more difficulty understanding spoken speech and more difficulty with the grammar of the language. Puberty seems to be an important time. An immigrant who speaks nothing but English from the age of eighteen on may still have a heavy accent in his old age; another immigrant who arrives at four years of age may have no trace of one.

Children who learn a second language when they are very young, between three and seven years of age, perform like native speakers on various tests. If they learn after eight years old, their performance declines gradually but consistently, especially during puberty. If you learn a second language after puberty, there is no longer any correlation between your age and your linguistic skill, which is consistently worse than it was when you were a child: twenty-year-olds do no better on the tests than forty-year-olds. This applies even to simple sound discriminations, like Pat's Japanese colleagues trying to hear and pronounce English *r*'s and *l*'s.

Why does it get so much harder to learn a new language as you get older? According to the classical critical-period argument, time is the important variable. If you haven't heard English *r*'s and *l*'s by the time puberty hits, you won't be able to discriminate them later, like the birds who couldn't learn songs after fifty days or the kittens who couldn't see after eighty days.

The alternative idea is that learning itself plays a role. As we saw earlier, babies who are exposed to a particular language early in infancy form prototypes—representations that describe that particular language's sound system. Developing these representations affects the babies' speech perception; the representations make some sounds indistinguishable from one another. Early in development we are open to learning the prototypes of many different languages. But by the time we reach puberty, these mental representations of sounds are well formed and become more fixed, and that makes it more difficult to perceive the distinctions of a foreign language. The representations we already have interfere with the representations of the new language.

If we adults don't learn as well because of this type of interference, then we might avoid it by creating a signal that gets by the interference. The studies of children with dyslexia we described earlier suggest that this might be true. School-age children with dyslexia have difficulty distinguishing sounds, such as *b* from *d,* and they have general language problems. Some studies indicate that these children improve if we help them separate sound categories. Children did better after they listened to computer-modified speech that exaggerated the differences between sounds. The exaggerated speech these dyslexic children hear in the treatment studies is in some ways like motherese, the speech grown-ups use when we speak to infants. This raises the possibility, for example, that we might help Japanese adults learn English by letting them listen to speech that exaggerates the differences between the categories *r* and *l*. That exaggerated speech might break past the interference of the Japanese representations these speakers have already developed. Even grown-ups might benefit from listening to motherese.

The Social Brain

One of the other surprises of recent studies on the brain's plasticity is that social factors can dramatically alter how animals learn. As we saw, white-crowned sparrows can typically learn their species' song from a tape recording between days twenty and fifty. However, this critical period seems less rigid in the right social context. The sparrows can learn after they are fifty days old if they are exposed to a live tutor, a real bird singing the song in front of them. Interacting with another bird helps the baby bird learn.

Some species of birds actually require social stimulation to learn. Zebra finches, for example, do not learn well from tape-recorded songs. They must interact with a tutor in a neighboring cage in order to learn his song. They won't learn if they just hear him. In fact, zebra finch fledglings can still learn from a tutor bird, even if they can't see him, as long as they interact with him in the usual ways that zebra finch fathers and sons do (finches peck and groom instead of playing catch, and they can find each other in the dark). A zebra finch will learn a "foreign" Bengalese finch song from a Bengalese finch foster father who feeds him, though he hears other adult zebra finch males singing the correct song nearby. Zebra finches seem, at least in part, to share the wisdom of Solomon: their dad is the one who shares their life, not just their genes.

What about babies? Do they have to interact with people to learn language? Could babies learn language just from a tape recorder or a television set? Clearly we can't separate babies from their parents and see what happens. But babies who are caught up in flirtatious dialogues with the people they love do seem very involved and attentive, and happy, too, and this could be part of why they learn so quickly.

The Brain in the Boat

The picture that emerges from the new brain research is consistent with the picture that emerges from the psychological research. Clearly, by the time babies are born, there is already a great deal of neurological structure in place. But, equally clearly, the brain changes in radical ways over the first few years of life, and it changes in response to experience. In other words, the brain learns. This learning isn't just passive. The brain actively tries to establish the right connections, and it prunes connections that don't get much use. The brain reprograms itself.

Moreover, the representations that result from learning influence how the brain processes new experiences. Experience changes the brain, but then those very changes alter the way new experience affects the brain. The sequence of development seems very important: choosing one path early on may heavily influence which paths will be available later.

Other people may also play a particularly important role in how the brain gets shaped. Even bird brains seem especially tuned to receive information from other birds, particularly nurturing birds. The very fact that so much of our human brain is devoted to processing language, and to understanding faces, suggests that for us, information from our "conspecifics" is even more important. In other words, the brain seems to love to learn from other people.

The brain evidence also supports the idea that babies and children have particularly powerful learning abilities and motivations, and that they do, in fact, learn more than we adults do. If you combine the psychological and neurological evidence, it is hard to avoid concluding that babies are just plain smarter than we are, at least if being smart means being able to learn something new. The advantage we adults have comes precisely from the fact that we once were babies. We can use the finely tuned, specialized, well-oiled mental machinery we constructed when we were very young to do all sorts of things that babies can't.

Andy and Alison once created a sci-fi fantasy about a world where we could somehow instantly give babies all the information scientists have and watch them ace our most difficult problems in six months or so. Or, in another fantasy, the great geniuses of science turn out to be the beneficiaries of a small mutation to their developmental timetable: they keep their baby brains a bit longer than the rest of us. Of course, these are just fantasies, but fantasies with a grain of truth. The instruments of culture—our ability to communicate and transfer one generation's discoveries to another—also help babies be such miraculous learners. And as adults we at least sometimes seem to retain our childlike ability to learn.

Although we may not be as smart as babies are, the new evidence suggests that we may be smarter than we sometimes think. The reason we don't learn more may be exactly because we have already learned so much. The wiring we acquired in childhood literally as well as metaphorically tells us most of what we need to know, it works staggeringly well most of the time, and we are designed in a way that makes those successful programs difficult to change. Even as adults, however, when we face new problems, unexpected environments, or unusual inputs, we seem to be able to change the wiring once more.

The brain itself, like knowledge, seems to be rather like Ulysses' boat, discussed in the previous chapter. It changes in a multitude of ways in response to the things that go on around it. Moreover, changes that take place early in the journey may dictate which changes will take place later on. If we decide to turn an oar into a mast, the oar won't be available to use as an anchor later. It seems that the biggest modifications are made early on; as the sailing gets smoother and the boat becomes more efficient, we need to change less and less. But we never really stop tinkering. Our brains are busy until the very end of the trip, when we and they inevitably become no more than inert slices on the neuroscientist's microscope stand. But even then, our knowledge boat becomes part of our legacy to our children, to rebuild and shape in their turn. Our brains, after all, have themselves been wired to theirs, even if the wiring uses songs and words and faces instead of electricity, and messages are transmitted through light and sound and touch instead of through synapses. Even after we are just slices on the stand, we still connect.

CHAPTER SEVEN

Trailing Clouds of Glory

Not in entire forgetfulness,
And not in utter nakedness,
But trailing clouds of glory do we come. . . .

—*Wordsworth*
"Ode: Intimations of Immortality from Recollections of Early Childhood"

What Is to Be Done?

Pat was on the telephone in March of 1997 when the secretary slipped her a note announcing: "The White House is on line #1." It was a conversation stopper, and she took the call. Mrs. Clinton's secretary was calling to ask Pat to speak at a White House conference in April 1997 called "Early Learning and the Brain." The issue of early development had become such a hot political topic that President and Mrs. Clinton had organized a conference. Six experts, including Pat, were to report the new research about brains and minds in the first three years of life. This conference generated acres of press and reams of images of cute babies on magazine covers and news shows. Requests for advice, reprints, and newspaper, magazine, and television interviews flooded Pat's office.

What should we tell the President? Or, for that matter, what should we tell the parents who called to ask how they should talk to their babies? One answer is the whole content of this book. There is a vast body of scientific work to report, and most of it is reasonably easy to understand. People who care about babies should be able to read about the science of development in places other than technical scientific journals or newspaper advice columns. But there are no simple translations of the science into policy decisions, either at the small scale of the parent grappling with the terrible twos or at the grand scale of the White House.

Raising children is an intrinsically difficult and uncertain job in ways that science can't really address. For most of us parents there is literally nothing more important than the well-being of our children. There are not many things we could imagine giving our lives for, but we could give our lives for them. And, in a less melodramatic way, of course, we do give our lives for them. For fifteen or twenty years our everyday energy, our individual liberty, our income, our attention, our concern are all devoted to our children. There is nothing else in human experiencc to match it.

And yet all this seriousness and commitment, this moral purpose, is combined with a deep, even necessary, lack of control. A British prime minister once intoned that the press wanted "power without responsibility, the prerogative of the harlot throughout the ages." Perhaps it's fitting that the prerogative of the mother is the opposite of the prerogative of the harlot: we parents have responsibility without power. Mothers and fathers are at the mercy of innumerable accidents—accidents about the random genetic mix of temperament and ability, accidents about how that mix interacts with our own temperaments and abilities, accidents about what our lives happen to be like in the few years that constitute a childhood, accidents about what the rest of the world has to offer our children.

There is also a deeper sense in which we have less control than responsibility. The whole point of the enterprise, after all, is to end up creating an autonomous agent, a person who can leave us, who can choose to make grave mistakes and decide to be thoroughly miserable. It's like falling utterly, madly, deeply in love and yet knowing that in twenty years the object of your affections will leave you for other lovers and, in fact, that your job is to make your beloved leave you for other lovers. The very best outcome is that our children will end up as decent, independent adults who will regard us with bemused and tolerant affection; for them to continue to treat us with the passionate attachment of infancy would be pathological. Almost every hard decision of child-rearing, each tiny step—Should I let her cross the street? Can he walk to school yet? Should I look in her dresser drawer?—is about how to give up control, not how to increase it; how to cede power, not how to gain it.

It is no surprise, then, that parents feel both a deep need for guidance and a deep ambivalence about it. We wish *someone* would tell us what to do, but on the other hand, we don't want anybody telling *us* what to do.

It is also not surprising that assorted quacks, con artists, and bullies have been happy to give advice, often invoking scientific authority. There is a largely dishonorable history of "expert" advice to mothers. To a developmental scientist, the most striking thing about most of this advice is how removed it was from any real empirical evidence or experimental research.

It is difficult to look at this history without seeing a classical feminist story, male "experts" ordering women around on the basis of a science that was supposed to be incomprehensible to them. But there is something else at work here, too. The very urgency and intractability of the problems lead to the development of pseudoscientific claims to expertise. The history of medicine provides a parallel. Illness is such an urgent concern for human beings that they have been willing to accept almost any kind of advice and therapy with almost any kind of invocation of knowledge and authority, from shamans' spells to bloodletting. In the past hundred years we have been starting to integrate real science into this enterprise, but it remains a very difficult task. We don't really care if our doctors appreciate the complex web of factors involved in our cancers or migraines or anxiety attacks; we just want them to get us well again.

Much the same factors influence the history of parenting advice. We want certainty, and that leaves us open to fraud. Mothers used to lie awake listening to their babies scream because the experts said not to pick the baby up or to feed "off schedule." They might well have felt that bloodletting would have been preferable.

One benefit of knowing the science is a kind of protective skepticism. It should make us deeply suspicious of any enterprise that offers a formula for making babies smarter or teaching them more, from flash cards to Mozart tapes to Better Baby Institutes. Everything we know about babies suggests that these artificial interventions are at best useless and at worst distractions from the normal interaction between grown-ups and babies. Babies are already as smart as they can be, they know what they need to know, and they are very effective and selective in getting the kinds of information they need. They are designed to learn about the real world that surrounds them, and they learn by playing with the things in that world, most of all by playing with the people who love them. Not the least advantage of knowing about science is that it immunizes us from pseudoscience.

Does science provide any more positive advice? The most important positive advice is that parents, and grown-ups in general, need to be allowed the time and energy to exercise their natural ability to help babies learn. One thing that science tells us is that nature has designed us to teach babies, as much as it has designed babies to learn. Almost all of the adult actions we've described are swift, spontaneous, automatic, and unpremeditated. And for babies and young children, care and teaching are inseparable. The very same actions that nurture babies give them the kinds of information they need. Combining our instinctive responses to babies and a long tradition of practical observation and experiment is likely to be the best way to decide what to do. The best experts, a Berry Brazelton or a Benjamin Spock or a Penelope Leach, actually provide just that sort of combination. The scientific research says that we should do just what we do when we are with our babies—talk, play, make funny faces, pay attention. We just need time to do it.

The science also suggests, however, that neither babies nor adults have some set of fixed, reflexive ways of learning and teaching, set down by evolution in the Pleistocene and forever more inalterable. On the contrary, the flexibility of the human cognitive system is as impressive as its initial richness. We are creatures born to learn about new places and to adjust to what we find there. There is certainly what we might think of as the "butter effect"—the fact that we still have responses, such as a passion for animal fat, that served us well in the Pleistocene but are maladaptive now. But an even stronger instinct leads to what we might think of as the "stationary-bike effect." Humans love explanations as much as we love piecrust and shortbread. Our explanatory drive can help us discover new things about ourselves. We can discover our attraction to animal fat and its perils and invent new tools to deal with it. We can remake our world, or at least add stationary bikes to it. This response is just as natural as the response to butter itself.

This is true about raising children, too. People have always had to adjust their child-rearing to suit the environments in which they found themselves. In tragic circumstances, where there were not enough resources to go around, people have even sometimes had to learn not to nurture. In evil times parents may even systematically abandon some children, though this always seems to come at great emotional cost. Anthropologists sometimes argue that this means there is no natural, innate nurturing capacity. But this is only true if we think nature equipped us with nothing more than fixed, inflexible ways of acting. In fact, the research suggests that nurturing is a natural inclination, but the way we act on it may be very different in different environments.

One way of thinking about our present dilemma is that our child-rearing environment has radically changed, but we have not yet worked out how to change what we do in response. A hundred years ago there were not only relatively few American mothers working outside the home, there were almost equally few American fathers working outside the home. This is because the vast majority of Americans were farmers, and farm families, men, women, and children, work and live, learn and teach in the same place. The movement of women out of the home and into the workplace parallels the movement of men out of the home and into the workplace, with only a relatively small historical lag. (Politicians notoriously have little historical sense, and that may be why we don't hear calls for men to give up their office jobs and go back to the home.) But this change in our environment, the change to an industrial rather than an agricultural or hunter-gatherer economy, obviously has important consequences for child-rearing.

Economists often tell us that family income hasn't declined since 1973. But that neglects the enormous fact that both parents work—and often work harder and longer hours—to provide the same income that one parent provided thirty years ago. Children, in particular, have suffered a grievous decline in just the goods that are most important to them: adult time, energy, and company. The child-rearing work that men and women and an extended family did a hundred years ago, and that women did thirty years ago, has to be done somehow by someone. The scientific moral is not that we need experts to tell us what to do with our children. What we need are the time and space and opportunity to do what we would do anyway, and that's just what we are losing.

The optimistic part is that, if we are right, humans are eventually able to alter their behavior and their cultural traditions in the light of new circumstances. We can continue to provide babies and young children with the rich physical and social environments that let them exercise their own spontaneous learning abilities, even in a postindustrial age. The invention of public schools was just such a response to changing circumstances. We're not going to go back to the farm anytime soon, and we have already exchanged the dark, satanic mills of industrialization for the fluorescently lit, though perhaps equally hellish, cubicles of the information economy. What we have to do is figure out how to raise our children here and now. What we need is the inventive intelligence and the will to make sure that babies and young children can exercise their innate ability to learn and that adults, not just mothers but all of us, can exercise our equally innate ability to teach them.

We need to do this both at a national level and in individual homes and lives, in states and cities and universities and corporations. An immediate step is to provide public support for care for infants and preschoolers. We provide publicly supported schools for older children because we know that educating children is a public good. And we provide publicly supported Social Security and Medicare because we know that at some times in our lives we are particularly vulnerable and in need of care. Both those arguments apply to care for babies and young children. Being in the company of caring adults *is* school for babies. This public support could take many forms, both as support for child-care centers with well-paid staff and as realistic subsidies or tax credits to parents. There might be a system of vouchers, for example, that parents could use both to pay for child care and to buy off time at work.

We could also immediately change workplaces to allow for part-time work that has similar benefits and pay to full-time work and to allow for flexible hours and career paths. Our own workplaces, the universities, provide both very good and very bad examples. For years professors have worked at home and determined their own schedules with no loss of productivity. On the other hand, the career structure of universities is deeply in conflict with the imperatives of evolution—the years when we expect academics to work the hardest and longest hours are exactly the years when women can have children.

The very automaticity of our response to babies suggests that it can be combined with doing other things, as it surely was in the Pleistocene. Perhaps the telecommuting home office with the crib next to the fax machine will turn out to be the contemporary equivalent of the baby in the sling on his mother's back or the father plowing next to his children. Perhaps the circle of fellow workers and friends will help replace the extended family group. Grandparents and uncles and aunts have also disappeared from children's lives just when they are most needed, and grandchildren and nieces and nephews have sadly disappeared from our lives. Perhaps we will construct institutions that allow people whose own children have grown up, or who don't have children, to be involved with other people's children.

Developmental science can play a role in giving us the information we need to redesign our tools. But if we are right, the capacity to think up new possibilities such as these, to explain, reason, and change, is not just the province of scientists, developmental or otherwise. The best way of integrating science and policy is to have a scientifically well-educated citizenry. Women, in particular, have learned that the best way of being responsible for their own health is to understand what biological science says and doesn't say, whether the topic is breast cancer or estrogen or childbirth, and then make their own decisions. Part of what we have tried to do in this book is to let nonscientists understand developmental science. In fact, an interesting conclusion of the developmental research we've talked about in this book is precisely that nonscientists *can* understand science, that science is an extension of our normal understanding of the world. We always have to make decisions based on what we know, and science is part of that knowledge. Knowing about developmental science may help us make informed decisions, but the decisions themselves are up to us.

The Clouds

One reason we care about babies is that we are responsible for what happens to them. But another, equally important reason to care about development is not that babies will turn into grown-ups but that we were once babies. When we study children, we are studying ourselves; when we see how they develop, we are seeing how we became what we are. The developmental research helps provide new empirical solutions to ancient philosophical problems, such as the problem of knowledge. We can understand how we come to know that other people have minds, that there is a world outside of us, that sounds have meanings. We can see how our knowledge emerges from the ideas we start out with, our ability to learn, and our interactions with other people. The picture of knowledge that emerges from looking at development turns out to apply quite broadly. It not only helps us understand what babies do, it also helps us understand what scientists do. It may even help us understand what artists and statesmen do.

We can also use babies, the best learners in the universe, as a model for other kinds of learning. The metaphor of the computer has helped us understand babies, but understanding babies will also allow us to construct new and more powerful kinds of computers. Babies may help us create machines that don't just implement predetermined programs but that actually interact with the outside world and learn from it. Understanding how babies decode speech is already helping us construct computers that can understand what we say to them. Understanding how babies learn helps us understand how we learn, but it also helps us understand how learning is possible. It helps us understand how any physical system could learn, including both the computers on our desks and the ones in our skulls.

But in the end, the real reason for studying babies and young children is just that they are themselves intrinsically so valuable and so interesting. When we look attentively, carefully, and thoughtfully at the things around us, they invariably turn out to be more interesting, more orderly, more complex, more strange, and more wonderful than we would ever have imagined. That's what happened when Kepler looked carefully at the stars, when Darwin looked at finches, when Marie Curie looked at pitchblende ore. And it's also what happened when Jane Austen looked at a provincial village and Proust looked at a madeleine cookie, when Vermeer looked at a girl making lace and Juan Gris looked at a café table.

That's what has happened as developmental psychologists have looked attentively, carefully, and thoughtfully at the minds of young babies and children. There is no great chain of knowing, beginning with simple and stupid babies, gradually advancing through childhood to ordinary adults, and reaching a peak in the geniuses of art and science. Babies' minds are at least as rich, as abstract, as complex, as powerful as ours. Babies think, reason, learn, and know as well as act and feel. At the same time, what they think is often radically different from what we think. Children are both surprisingly like us and surprisingly unlike us.

There are moral implications to this new view of babies and young children. The sort of policy discussions we described earlier are geared to the question of how we can turn babies and children into the right sort of grown-ups. But the new research shows that babies and young children are fully human beings in their own right. We may not have much control over how children turn out, but we do have enormous power over their lives as children, and those lives are as valuable and important as adult lives. Children aren't just valuable because they will turn into grown-ups but because they are thinking, feeling, individual people themselves.

Until very recently doctors didn't use analgesia when they operated on small babies, because they thought their minds were too primitive to really feel pain or to remember it if they did. This is a dramatic example, but it often seems as if we discount children's pain compared with adult pain. Child abuse isn't evil because it may produce neurotic adults but because it abuses children. Divorce doesn't have a cost because it may produce adults who have difficulty with relationships but because it causes emotional pain to children. Parents aren't important because they may shape their children's adult personalities but because they are the most profound influence on children's lives while they are children. Looking at babies attentively makes us treat them differently.

The deepest insight that comes out of looking attentively at babies, though, is understanding where our ability to look attentively comes from. The most interesting thing about babies is that they are so enormously interested; the most wonderful thing about them is their infinite capacity for wonder.

The distinguished developmental psychologist John Flavell once told us that he would trade all of his awards for the chance to see through a child's eyes for just a few minutes. The early nineteenth-century Romantic poets such as Wordsworth and Blake had the same ambition. They thought childhood was the time when we saw the universe most clearly and experienced it most intensely. It was "a time when meadow, grove, and stream, the earth, and every common sight, to me did seem apparelled in celestial light." The time when we saw "a world in a grain of sand and a heaven in a wild flower." Wordsworth and Blake also saw that even as adults we continue to have moments of this clarity and intensity of understanding. And they saw that those glimpses were part of the experience of creation: they let us write poetry.

We think Wordsworth and Blake were quite right about that. They were wrong, though, in thinking that those experiences were in opposition to the experience of reasoning, considering, deducing, and experimenting—in short, doing science. They were wrong to think that children's knowledge is the opposite of scientific knowledge. Blake made Newton into his great symbolic enemy, but, in fact, they were far more similar than they were different (right down to being occasionally nutty as fruitcakes). Science is not the cool, detached business of popular legend but much the same ecstatic enterprise as Romantic poetry (right down to the late nights and rumpled hairdos). At their best moments scientists also see the splendor in the grass, the glory in the flower. They also have glimpses of a kind of clarity, a particular combination of truth and beauty.

We hope our view of babies is not too romantic, but it is certainly Romantic. We think "Intimations of Immortality" and "Auguries of Innocence" may, in fact, describe just what it is like to be a child. That's the way the world looked when your brain had 15,000 synapses per neuron and burned up twice as much glucose. And that is still, at least sometimes, what it's like to be human, even if it takes the form only of intimations and auguries. It is, in particular, what it's like to create art or science, or to appreciate the art and science of others.

The new research says something else, too. It isn't just that we have these intimations of truth, this experience of understanding, this feeling that we are in tune with the world. We really do approach the truth, we really do understand, we really are in tune with the world. Nature has designed us to understand nature. Our eyes have evolved to give us an accurate picture of the world, and so have our brains. But, unlike our eyes, our brains don't give us just one answer to the question and stop there. Instead, we are designed to create a sequence of pictures of the world and perpetually to revise the pictures we currently have. That's what we do in science and art, and that's what we do practically all the time when we are very young.

Part of the Romantic sensibility, a part we inevitably share at least a little, was to grieve over the loss of this childlike clarity and its replacement by the more mundane duties and obligations of adult life. Getting and spending, we lay waste our powers; the things which we have seen, we now can see no more. It may seem that the Romantic view we are articulating sees ordinary adulthood as a loss, a falling off, only briefly stemmed by a few adult geniuses.

But that neglects the other half of the equation, the part that is our uniquely adult gift. In particular, when we take on the adult obligation of caring for children, we don't give up the Romantic project, we participate in it. We participate simply by watching children. Think of some completely ordinary, boring, everyday walk, the couple of blocks to the local 7-Eleven store. Taking that same walk with a two-year-old is like going to get a quart of milk with William Blake. The mundane street becomes a sort of circus. There are gates, gates that open one way and not another and that will swing back and forth if you push them just the right way. There are small walls you can walk on, very carefully. There are sewer lids that have fascinatingly regular patterns, and scraps of brightly colored pizza-delivery flyers. There are intriguing strangers to examine carefully from behind a protective parental leg. There is a veritable zoo of creatures, from tiny pill bugs and earthworms to the enormous excitement, or terror, of a real barking dog. The trip to the 7-Eleven becomes a hundred times more interesting, even though, of course, it does take ten times as long. Watching children awakens our own continuing capacities for wonder and knowledge.

But we are more than just witnesses to Romantic genius at work. When we take care of children, we are also helping the human species find the truth and understand the world. Of course, a lot of it is changing diapers and wiping noses and making peanut-butter sandwiches. And a lot of it is worry and exhaustion. But a lot of it, and a lot of the very best of it, the kisses and the pet names, the games and the jokes, turns out to be part of this larger enterprise. We might not have thought that flirting with babies helped solve the Other Minds problem, or that playing hide-and-seek had anything to do with metaphysics, or that baby talk held the answer to the problem of meaning. But that's just what developmental cognitive science has discovered. We human beings seem designed to complete our grandest projects by pursuing ordinary little joys.

NOTES

Chapter One: Ancient Questions and a Young Science

9 Nature and uses of immaturity: Bruner, 1972; Gould, 1977.

10 The *Meno:* Plato, 1937a.

12 The "great chain of knowing": Locke, 1959; Rousseau, 1974; Itard, 1962.

12 Mind as a "tabula rasa": Locke, 1959.

12 Wordsworth, 1943.

13 Views of children through the ages: Ariès, 1962.

14 Aristotle's powers of observation: Russell, 1948.

14 Piaget's autobiography: Piaget, 1952a. His classic books on early mental development: Piaget, 1952b, 1954, 1962. Analysis of his theory: Flavell, 1963.

17 Vygotsky's biography and work: Wertsch, 1985; Vygotsky, 1986.

19 Freud and Skinner: Freud's views did inspire other researchers to adopt experimental approaches to studying babies and young children: Bowlby, 1969, 1973; Winnicott, 1971; Ainsworth et al., 1978; Stern, 1985. Similarly, Skinner's work on conditioning laid the groundwork for experimental techniques currently used to study babies, as discussed in chapters 2–4.

21 The new field of cognitive science: A highly readable and acces-

sible account of the birth of the field can be found in Gardner, 1985.

Chapter Two: What Children Learn About People

23 Evolutionary significance of understanding of other minds: Byrne and Whiten, 1988; Whiten, 1991; Povinelli and Preuss, 1995; Povinelli and Eddy, 1996; Tomasello and Call, 1997.

27 Techniques for discovering what babies discriminate and prefer: Fantz, 1963; Kagan, 1970; Gottlieb and Krasnegor, 1985; Mehler and Fox, 1985; Gibson, 1987; Salapatek and Cohen, 1987; Kellman and Arterberry, 1998.

27 Babies get bored, "habituate": Cohen, 1979; Slater, Morison, and Rose, 1984; Hood et al., 1996; Quinn and Eimas, 1996; see also previous note.

27 Newborn preferences for Mother's voice and language: DeCasper and Fifer, 1980; Mehler et al., 1988; Moon, Cooper, and Fifer, 1993. Mother's smell: Macfarlane, 1975; Porter et al., 1991. Mother's face: Field et al., 1984; Bushnell, Sai, and Mullin, 1989; Walton, Bower, and Bower, 1992; Pascalis et al., 1995.

28 Babies' understanding of emotional expressions: Nelson, 1987; Walker-Andrews, 1997.

28 Knowing how people move: Bertenthal et al., 1985; Bertenthal, Proffitt, and Kramer, 1987; Bertenthal, 1993.

28 The development of vision in babies: Banks, 1980; Atkinson, 1984; Aslin, 1987; Kellman and Banks, 1998.

29 Imitation of facial gestures: Meltzoff and Moore, 1977, 1983, 1992, 1994. For cross-cultural extensions, analyses of how and why babies copy others, and what babies learn from imitation, see Meltzoff and Moore, 1997, and Nadel and Butterworth, 1999.

30 Babies' jump start on the Other Minds problem: Meltzoff and Gopnik, 1993; Gopnik and Meltzoff, 1994; Meltzoff and Moore, 1995.

30 Babies are tuned in to people: Watson, 1972; Brazelton, Koslowski, and Main, 1974; Bruner, 1975, 1983; Trevarthen, 1979; Brazelton and Tronick, 1980; Stern, 1985; Muir and Hains, 1993; Bråten, 1999.

32 A striking change in interacting with people: Trevarthen and Hubley, 1978; Campos and Stenberg, 1981; Baldwin and Moses, 1994; Carpenter, Nagell, and Tomasello, 1998.

32 Understanding pointing: Desrochers, Morissette, and Ricard, 1995; Franco and Butterworth, 1996; Butterworth, 1997; O'Neill, 1996.

33 Looking at one box with disgust and another with delight: Repacholi, 1998.

33 Babies learn how to use novel objects by watching adults: Meltzoff, 1988a, b; Klein and Meltzoff, 1999. They also learn by watching adults on TV: Meltzoff, 1988c.

35 Young children as cultural beings: Meltzoff, 1988d; Bruner, 1990; Rogoff, 1990; Tomasello, Kruger, and Ratner, 1993; Meltzoff and Moore, 1999a.

36 Conflicting desires as investigated by delicious Goldfish crackers vs. yucky broccoli: Repacholi and Gopnik, 1997.

37 "Terrible twos": Gopnik and Meltzoff, 1997; Repacholi and Gopnik, 1997.

39 Empathy: Brothers, 1989; Harris, 1989; Zahn-Waxler et al., 1992.

40 Changing your point of view and taking others' perspective: Masangkay et al., 1974; Lempers, Flavell, and Flavell, 1977; Flavell et al., 1981; Gopnik, Slaughter, and Meltzoff, 1994; Gopnik, Esterly, and Meltzoff, 1995.

42 The conversational attic and CHILDES: MacWhinney and Snow, 1985, 1990. Analyses of early uses of mental state terms and examples are taken from Bartsch and Wellman, 1995.

44 About "aboutness": Brentano, 1973; Frege, 1952; Flavell et al., 1981; Leslie, 1987; Forguson and Gopnik, 1988; Wellman, 1990; Perner, 1991.

44–45 Sir Walter Scott, candy, trick boxes, and false beliefs: Wimmer and Perner, 1983; Flavell, Green, and Flavell, 1986; Perner, Leekam, and Wimmer, 1987; Astington, Harris, and Olson, 1988; Gopnik and Astington, 1988; Gopnik and Slaughter, 1991; Astington, 1993; Taylor, 1996; Flavell and Miller, 1998.

46 Children can't remember the source of their knowledge: Gopnik and Graf, 1988; Wimmer, Hogrefe, and Perner, 1988; O'Neill and Gopnik, 1991; O'Neill, Astington, and Flavell, 1992.

46 Even young children can remember for a long time: Meltzoff, 1995b.

48 Freud's passionate babies: Freud, 1953.

48 Lorenz's goslings: Lorenz, 1937.

48 Mother-infant bonding: Klaus and Kennell, 1982; Klaus, 1998.

49 Internal working models, theories of attachment, and what love has to do with it: Bretherton and Waters, 1985; Main, 1991; Waters et al., 1995; Werner and Smith, 1998.

50 Children are bad liars: Chandler, Fritz, and Hala, 1989; Sodian,

1991; Sodian et al., 1991; Peskin, 1992; Russell, Jarrold, and Potel, 1994.

51 Politeness and masking your emotions: Harris, 1989.

51 Why formal schooling starts at five to six years of age: Rogoff et al., 1975; Gardner, 1991; Taylor, Esbensen, and Bennet, 1994; Astington and Pelletier, 1996; Bruner, 1996.

52 Childhood amnesia: Nelson, 1990; Gopnik, 1993; Newcombe and Fox, 1994; Meltzoff, 1995b; Perner and Ruffman, 1995.

53 Mind-blindness in autism: Hobson, 1993; Baron-Cohen, 1995; Grandin, 1995; Happé, 1995; Sacks, 1995; Sigman and Capps, 1997; Dawson et al., 1998; Baron-Cohen, Tager-Flusberg, and Cohen, 1993, 1999.

55 Becoming a psychologist: Wellman, 1990; Perner, 1991; Gopnik, 1993; Gopnik and Wellman, 1994; Bartsch and Wellman, 1995; Flavell and Miller, 1998; Meltzoff, Gopnik, and Repacholi, 1999.

56 Children learn like scientists: Slaughter and Gopnik, 1996.

57 Differences between sibs in personality and IQ: Dunn and Plomin, 1990; Sulloway, 1996.

58 Older siblings influence the development of younger ones: Perner, Ruffman, and Leekam, 1994; Brown, Donelan-McCall, and Dunn, 1996; Jenkins and Astington, 1996; Ruffman et al., 1998.

Chapter Three: What Children Learn About Things

61 Magic shows and what it's like to be a baby: Gopnik and Meltzoff, 1997.

62 Tortuous path from world to brain: Aristotle, 1956; Descartes, 1952; Berkeley, 1910.

62 Modern answer: Marr, 1982; Pinker, 1997; Palmer, 1999.

64 The allure of stripes: Haith, 1980.

65 Importance of movement and common fate: Bower, 1982; Kellman and Spelke, 1983; Slater, Morison, and Rose, 1984; Hofsten and Spelke, 1985; Spelke et al., 1993; Kellman and Arterberry, 1998.

66 Predicting where objects will appear: Moore, Borton, and Darby, 1978; Bower, 1982; Baillargeon and Graber, 1987; Spelke et al., 1992; Munakata et al., 1997; Haith, 1998; Meltzoff and Moore, 1998.

67 Berkeley thought that touch teaches vision: Berkeley, 1910.

67 Distance and 3-D perception in young babies: Bower, 1982; Yonas and Owsley, 1987.

68 Size constancy: Bower, 1982; Granrud, 1987; Slater, Mattock, and Brown, 1990.

68 Locke's problem of a blind man who gains sight: Locke, 1959. See Sacks, 1995, for a modern example of surgically removing cataracts from a blind person and Meltzoff, 1990b, for a discussion of coordinating sight and touch in normal development.

69 Using pacifiers to solve Locke's problem: Meltzoff and Borton, 1979; Gibson and Walker, 1984; Kaye and Bower, 1994. Also see Bryant et al., 1972, for work with slightly older babies.

69 Auditory-visual correspondences: Spatial location: Wertheimer, 1961; Morrongiello, 1994. Temporal synchrony: Spelke, 1979, 1987; Bahrick, 1987; Lewkowicz and Lickliter, 1994.

69 "*Ahhhhhh!*"—lipreading in babies: Kuhl and Meltzoff, 1982, 1984; see also MacKain et al., 1983; Walton and Bower, 1993.

71 Object disappearances in the first six months of life: There is considerable debate concerning what young babies understand about objects that are occluded: Bower, 1982; Baillargeon, 1993; Gopnik and Meltzoff, 1997; Munakata et al., 1997; Haith, 1998; Meltzoff and Moore, 1998, 1999b; Spelke, 1998.

72 Recovering hidden objects in eight- to twenty-four-months-olds: Piaget, 1954; Bower, 1982; Butterworth and Jarrett, 1982; Harris, 1987; Moore and Meltzoff, 1999.

73 Hume's causality: Hume, 1984.

74 Young babies detect contingencies between their actions and events: Lipsitt, 1969; Lipsitt and Werner, 1981; Papousek, 1969; Papousek and Papousek, 1984; Watson, 1972; Rovee-Collier and Lipsitt, 1982; Bower, 1989.

74 Tying ribbons on babies: Rovee-Collier and Gekoski, 1979; Rovee-Collier et al., 1980; Rovee-Collier, 1990.

75 Magical and irrational thinking in children and adults: Piaget, 1954; Evans-Pritchard, 1976; Atran, 1990; Shweder, 1991; Boyer, 1994; D'Andrade, 1995; Cole, 1996; Shore, 1996; Sperber, 1996; Shweder et al., 1998; Lillard, 1998.

76 Cloth pulling and causality: Willatts, 1984, 1989.

77 Using rakes as tools: Piaget, 1954; Uzgiris and Hunt, 1975; Gopnik and Meltzoff, 1986.

78 Billiard-ball causality: Leslie, 1984; Leslie and Keeble, 1987; Oakes and Cohen, 1995. Leslie says six-month-olds know about billiard ball–type causal relations, whereas Oakes and Cohen say this first emerges at about ten months of age. For studies of the roots of

psychological causality, see Gergely et al., 1995; Meltzoff, 1995a; Woodward, 1998.

78 Children continue to learn and begin to offer causal explanations: Bullock and Gelman, 1979; Gelman, Bullock, and Meck, 1980; Kalish, 1988; Wellman, Hickling, and Schult, 1997.

79 Platonic love, sweet-peas, and kinds of things: Plato, 1937b, 1951.

79 Categorization based on similarity: Kripke, 1972; Goodman, 1983.

80 Object essences and categorization: Putnam, 1975; Kripke, 1980; Keil, 1989; Gelman and Wellman, 1991; Wellman and Gelman, 1992, 1998.

81 Magical object switches: Moore, Borton, and Darby, 1978; Bower, 1982; Xu and Carey, 1996; Meltzoff and Moore, 1998.

81 New understanding of categories: Ricciuti, 1965; Sugarman, 1983; Gopnik and Meltzoff, 1987, 1992; Mervis and Bertrand, 1994.

82 Rhinoceros and triceratops: Gelman and Markman, 1987; Markman, 1989; Gelman and Coley, 1991.

82 Blicket detectors: Gopnik and Sobel, 1997.

83 Outsides vs. insides: Springer and Keil, 1989, 1991; Gelman and Wellman, 1991; Springer, 1996.

83 Early understanding of biology: Springer and Keil, 1989, 1991; Wellman and Gelman, 1992; Hickling and Gelman, 1995; Springer, 1996.

84 Williams syndrome: Bellugi, Wang, and Jernigan, 1994; Johnson and Carey, 1998; Mervis and Bertrand, 1997.

85 The explanatory drive: Gopnik and Meltzoff, 1997; Gopnik, 1998; Keil and Wilson, 1998.

88 The mooing can: Baldwin, Markman, and Melartin, 1993.

88 Grown-ups as teachers: Bruner, 1983, 1996; Rogoff, 1990, 1998; Nelson, 1996.

89 Korean- and English-speaking parents: Gopnik and Choi, 1990; Choi and Gopnik, 1995; Gopnik, Choi, and Baumberger, 1996; Tardif, Shatz, and Naigles, 1997; Gelman and Tardif, 1998.

90 Whorfian hypothesis: Carroll, 1956.

Chapter Four: What Children Learn About Language

92 Seventy-five thousand words and an infinite number of combinations: Chomsky, 1980; Levelt, 1989; Pinker, 1994; Miller, 1996.

94 Speech as a cryptogram: Liberman et al., 1967.

94 Having a conversation with a computer: "Mr. Gates builds his brain

trust," *Fortune,* Dec. 8, 1997, 84–98; "Let's talk! Speech technology is the next big thing in computing," *Business Week,* Feb. 12, 1998, 60–72.

95 Bill Gates on computer speech understanding: "Microsoft: Beyond talking Barney. It's tedious work, but the software giant dearly wants PCs to gab," *Business Week,* Feb. 12, 1998, 80.

95 Spectrograms of speech: Stevens, 1998.

95 Different voices, rates, and contexts—why speech is so difficult to decode: Kuhl, 1994a.

96 The sounds used in human languages: Ladefoged and Maddieson, 1996; Crystal, 1997.

96 Talking . . . very . . . slowly . . . to a computer: DragonDictate was one of the early speech recognition programs that was effective with slow speech. Bamberg and Mandel, 1991.

97 Continuous speech recognition by computer: Kurzweil, 1999, describes the new software used for dictation. He also makes interesting predictions about the capabilities of future computers.

98 Saint Augustine's solution to naming: Saint Augustine, 1995.

98 Bertrand Russell's imaginary objects: Russell, 1905.

98 Wittgenstein's intentions: Wittgenstein, 1953.

98 Quine's spaces: Quine, 1960.

99 Chomsky's revolution: Chomsky, 1957, 1965, 1980.

100 Language evolution: Bickerton, 1981, 1990, 1995; Kuhl, 1988; Lieberman, 1991; Pinker, 1994; Hauser, 1996; Deacon, 1997.

100 Where the linguistic system comes from: Brown, 1973; Bates et al., 1979; Bruner, 1983; Markman, 1989; Bloom, 1993; Hirsh-Pasek and Golinkoff, 1996; Nelson, 1996; Gopnik and Meltzoff, 1997.

101 Different communities speak different languages: Slobin, 1992–1997.

102 Languages carve up sounds differently: Lisker and Abramson, 1964, for Thai; Abramson and Lisker, 1970, for Spanish; Miyawaki et al., 1975, for Japanese.

103 Categorical perception: Studdert-Kennedy et al., 1970.

103 Japanese listeners' discrimination of American *r* and *l:* Goto, 1971; Strange and Dittmann, 1984; Logan, Lively, and Pisoni, 1991.

104 What differs is our brains: Kuhl, 1994b.

104 Techniques for testing babies under four months old: Jusczyk, 1985.

105 Babies' speech discrimination: Kuhl, 1987; Jusczyk, 1997.

106 Kikuyu, French, or Chinese, babies are "citizens of the world":

Eimas et al., 1971; Eimas, 1975; Lasky, Syrdal-Lasky, and Klein, 1975; Streeter, 1976.

106 Babies make distinctions no matter who is talking: Kuhl, 1985a.

107 Techniques for testing six- to twelve-month-olds: Kuhl, 1985b.

108 What happens between six and twelve months?: Werker and Tees, 1984; Werker, 1991; Kuhl et al., 1992; Kuhl, 1998. For speech discrimination at fourteen months: Stager and Werker, 1997.

108 Prototypical sounds: Kuhl, 1991, 1994b; Kuhl and Iverson, 1995.

108 Categories and prototypes can act like filters and can distort perception: Rips, 1975; Rosch, 1975; Mervis and Pani, 1980; Mervis and Rosch, 1981; Weber and Crocker, 1983; Miller, 1994; Nygaard and Pisoni, 1995; Iverson and Kuhl, 1995; Kuhl, 1998.

109 Finding words in the stream of speech: Cutler and Butterfield, 1992; McQueen, Norris, and Cutler, 1994.

109 Learning the regularities of your native language: Jusczyk, Cutler, and Redanz, 1993; Jusczyk et al., 1993; Saffran, Aslin, and Newport, 1996.

111 Babbling babies: Ferguson, Menn, and Stoel-Gammon, 1992; Kent, 1992; Locke, 1993. Deaf babies babble with their hands: Petitto, 1993.

112 Why use *Mama* and *Dada*?: Murdock, 1959; Jakobson, 1960.

113 *Gone, there,* and other nonobject words: Bloom, 1973; Gopnik, 1982, 1984, 1988b; Nelson, 1985; Tomasello and Merriman, 1995.

115 When the mailman is "daddy": Clark, 1974; Bowerman, 1989; Naigles and Gelman, 1995.

115 Babies suddenly name everything in sight: Nelson, 1985; Reznik and Goldfield, 1992; Fenson et al., 1994; Woodward, Markman, and Fitzsimmons, 1994.

116 Fast mapping: Carey, 1978; Mervis and Bertrand, 1994.

116 Language is as much invented as learned: Mervis, 1987.

116 Using assumptions to decode language: Markman, 1989; Golinkoff, Mervis, and Hirsh-Pasek, 1994.

116 Reading the intentions of other people: Tomasello and Barton, 1994; Meltzoff, 1995a; Meltzoff, Gopnik, and Repacholi, 1999; Baldwin et al., 1996; Tomasello, Strosberg, and Akhtar, 1996.

116 Apples, pears, and "daxes": Baldwin, 1993a, b; Baldwin et al., 1996.

117 Putting words together: Brown, 1973; Bates, Bretherton, and Snyder, 1987; Bloom, 1991; Fletcher and MacWhinney, 1995; Hirsh-Pasek and Golinkoff, 1996.

118 Klingon talkers: Nelson, 1981.

119 Learning plurals: Berko, 1958; Mervis and Johnson, 1991.

119 Children learning different languages: Slobin, 1992–1997.

121 Children with dyslexia: Tallal, Miller, and Fitch, 1993; Studdert-Kennedy and Mody, 1995.

121 Genetically determined language disorders: M. Gopnik, 1990; M. Gopnik and Crago, 1991.

122 Babies hear hundreds of thousands of vowels: Chapman et al., 1992; Kuhl, 1994b.

122 Abstracting mental prototypes: Posner and Keele, 1970; Goldman and Homa, 1977; Strauss, 1979; Bomba and Siqueland, 1983; Medin and Barsalou, 1987; Kuhl, 1991, 1994b; Estes, 1993; Nygaard and Pisoni, 1995; Iverson and Kuhl, 1996. Prototype effects occur for speech, faces, and dot patterns; they could be due either to abstract summary representations babies form after exposure to many instances or to memory traces of the instances themselves.

124 Learning to produce the sounds of your native language: Kent, 1992; Oller and Lynch, 1992; de Boysson-Bardies, 1993; Vihman and de Boysson-Bardies, 1994.

124 Babies talk back, imitating adults: Kuhl and Meltzoff, 1996.

125–27 Relations between language and thought in toddlers: *Gone* and object permanence: Gopnik, 1984; Gopnik and Meltzoff, 1984, 1986. *Uh-oh* and tool use: Gopnik and Meltzoff, 1986. Names of objects and categories: Gopnik and Meltzoff, 1987, 1992. For a theory of relating language and thought in early childhood, see Gopnik and Meltzoff, 1997. For related work, see also Bloom, 1973; Tomasello and Farrar, 1986; Lifter and Bloom, 1989.

128 Sweet sounds of motherese: Stern et al., 1983; Fernald and Simon, 1984; Fernald, 1992; Aslin, 1993.

128 Fathers speak motherese, too: Jacobson et al., 1983.

128 Babies love motherese: Fernald, 1985; Fernald and Kuhl, 1987.

129 Motherese is universal: Ferguson, 1964; Blount and Padgug, 1977; Grieser and Kuhl, 1988; Fernald et al., 1989.

130 Shorter and simpler sentences: Ferguson, 1964; Snow, 1977; Snow and Ferguson, 1977.

130 Parents speak more clearly when using motherese: Kuhl et al., 1997.

131 Bootstrapping: Pinker, 1984, 1987; Morgan and Demuth, 1995.

Chapter Five: What Scientists Have Learned About Children's Minds

134 Descartes's answer: Descartes, 1911.

134 "The ghost in the machine": Ryle, 1949.

135 Computers lie between mind and desk, with some desklike and some mindlike qualities: Turing, 1950; Haugeland, 1989, 1997; Pinker, 1997. For opposing views: Searle, 1984; Dreyfus, 1992.

136 Programs, representations, and rules: Fodor, 1975; Dennett, 1978; Block, 1990.

137 Remarkable discoveries about computers: Ritchie, 1986; Herken, 1988.

137 Selected discussions of the basic idea of cognitive science: Haugeland, 1981, 1989; Pylyshyn, 1984; Pinker, 1997. The invention of new "connectionist" computer programs has led to a debate about what mental representations and rules are like: Clark, 1993; Haugeland, 1997.

137 Special types of programs: Video image translating to object descriptions: Hildreth and Ullman, 1989; Haralick and Shapiro, 1992; Pinker, 1997. Diagnose illness: Miller, Pople, and Myers, 1982; Middleton, Shwe, and Heckerman, 1991. Analyze Mars rocks: Glymour, Ramsey, and Roush, 1999.

138 Android epistemology: Ford, Glymour, and Hayes, 1995.

140 Consciousness and subjective "feel": Churchland, 1988; Dennett, 1991; Carruthers, 1996; Chalmers, 1996; Lycan, 1996.

141 Philosophers on children's thought: Davidson, 1980.

143 Zeitgeist change and the history of the International Conference on Infant Studies: Lipsitt, 1978, 1998.

148 Modern developmental scientists don't advocate extreme maturational or association views: For a sampling of current thinking, see Karmiloff-Smith, 1992; Kagan, 1998; Elman et al., 1996; Gopnik, 1996a, b; Nelson, 1996; Gopnik and Meltzoff, 1997; Pinker, 1997; Gelman and Williams, 1998; Kuhl, 1998; Siegler, 1998; Spelke and Newport, 1998; Wellman and Gelman, 1998; Flavell, 1999; Meltzoff and Moore, 1998, 1999b.

150 Ulysses' boat: Neurath, 1959. At a broad level this view is similar to Piaget's constructivism, though the substance of our theory, especially our views on the power of the initial representational system, is very different. See Gopnik and Meltzoff, 1997.

153 Play is education for babies: Piaget, 1962; Vygotsky, 1967; Bruner, 1973.

155 Theory theory: In philosophy: Morton, 1980; Stich, 1983; Church-

land, 1981, 1995. In psychology: Gopnik, 1988a; Karmiloff-Smith and Inhelder, 1974; Carey, 1985, 1988; Keil, 1989; Wellman, 1990; Gopnik and Wellman, 1994; Gopnik and Meltzoff, 1997.

156 Children and scientists as similar: Gopnik, 1996b; Gopnik and Meltzoff, 1997.

157 Cognitive immaturity, flexibility, and learning: Bruner, 1972; Bennett and Harvey, 1985; Bjorklund, 1997.

159 Folk botany and Australian aboriginal geography: Lewis, 1976; Atran, 1990.

160 The division of labor and organized science: Kitcher, 1993.

162 Orgasms of the mind: Gopnik, 1998.

162 Babies' emotions and negative reactions when they can't make sense of things: Moore and Meltzoff, 1999.

163 "Nature seems to act on us as a teaching machine": S. Weinberg, "The revolution that didn't happen," *The New York Review of Books*, Oct. 8, 1998.

163 "There is lust of the mind": Hobbes, 1962.

165 Parents and caretakers in the behaviorist tradition: Watson, 1928, 1930; Skinner, 1948, 1971.

165 Why do babies look so cute?: Lorenz, 1943; Fullard and Reiling, 1976.

166 Babies like adults who imitate them: Meltzoff, 1990a; Meltzoff and Moore, 1999a. They also like playing imitation games with peers: Nadel-Brulfert and Baudonnière, 1982.

167 Baby imitation as the cradle of culture: Meltzoff, 1988d; Meltzoff and Moore, 1999a; Tomasello, Kruger, and Ratner, 1993.

169 Social influence of peers, not just mothers: Babies learn from and imitate other children in day care: Hanna and Meltzoff, 1993. Even four-year-olds use motherese to speak to younger children: Shatz and Gelman, 1973.

170 Idealized, logical scientist: Hempel, 1965; Popper, 1965; Kitcher, 1993.

170 History and sociology of science: Kuhn, 1962.

170 Postmodern view of science: Feyerabend, 1975.

172 On genius and leadership: Gardner, 1995, 1997.

172 On politics: Habermas, 1979.

Chapter Six: What Scientists Have Learned About Children's Brains

174 Three pounds of gray jelly: Edelman, 1992; Kosslyn and Andersen, 1992; Shumeiko, 1998.

175 Mind depends on brain: P. S. Churchland, 1986; P. M. Churchland, 1995.

176 Aphasia, a language disorder: Caplan, 1992; Goodglass, 1993.

176 The specialized brain: Geschwind, 1979; Nass and Gazzaniga, 1985; Geschwind and Galaburda, 1987.

177 Individual brain cells fire in response to: Sounds: Morel, Garraghty, and Kaas, 1993. Faces: Desimone et al., 1984; Perrett, Mistlin, and Chitty, 1987; Perrett et al., 1992. Social signals: Perrett et al., 1990. Seeing and making movements: Rizzolatti et al., 1996; Rizzolatti and Arbib, 1998. Language: Ojemann, 1983.

178 Recording magnetoencephalographic (MEG) and event-related potentials (ERP) while listening to speech: Eulitz et al., 1996; Näätänen et al., 1997.

179 PET scans of the brain: Petersen et al., 1990; Zatorre et al., 1992; Petersen and Fiez, 1993; Morris et al., 1996.

179 Functional magnetic resonance imaging (fMRI) and language: Bavelier et al., 1997; Kim et al., 1997.

179 Brain mapping prior to brain surgery: Gallen et al., 1995; Ganslandt et al., 1997.

179 Lighting up the thinking brain: Damasio and Damasio, 1992; Posner and Raichle, 1994.

180 Building a brain from birth to six years: Shankle et al., 1998.

181 Wiring the brain: Shatz, 1992.

181 Brains of people with Alzheimer's disease: Frey, Minoshima, and Kuhl, 1998; Mielke and Heiss, 1998.

182 Rats living in "rich" environments: Diamond, Krech, and Rosenzweig, 1964; Greenough, Volkman, and Juraska, 1973.

183 Wiring the kitten's eye to the kitten's brain: Hubel and Wiesel, 1963, 1965, 1970.

184 "Cells that wire together fire together": Shatz, 1990.

185 Making connections between brain cells: Huttenlocher, 1979, 1990; Huttenlocher and de Courten, 1987; Chugani, 1998.

185 Growing synapses and dendrites: Simonds and Scheibel, 1989; Jacobs, Schall, and Sheibel, 1993; Jones et al., 1997.

186 Synaptic pruning: Chugani, Phelps, and Mazziotta, 1987; Chugani, 1994; Huttenlocher, 1994.

188 Babies' brains recognize native-language speech prototypes: Cheour et al., 1998.

189 Adult brains continue to make new connections and generate new

brain cells: Nottebohm, Nottebohm, and Crane, 1986; Kirn and Nottebohm, 1993; Eriksson et al., 1998; Gould et al., 1998.

190 Critical periods in animals: Nottebohm, 1969; Hubel and Wiesel, 1970; Marler, 1970a; Konishi, 1985; Brainard and Knudsen, 1998; Knudsen, 1998.

191 On critical periods for language learning: Dennis and Whitaker, 1976; Snow, 1987; Newport, 1990; Duchowny et al., 1996; Vargha-Khadem et al., 1997.

192 Genie and other "wild children": Fromkin et al., 1974; Lane, 1976; Curtiss, 1977.

192 Difficulty of second language learning after puberty: Johnson and Newport, 1991; Newport, 1991.

192 Speaking a foreign language with an accent: Oyama, 1976; Flege, 1988; Newport, 1991.

193 On parallels between critical periods in birds and babies: Marler, 1970b; Kuhl, 1989; Doupe and Kuhl, 1999.

193 Reexamining critical periods as due to prior learning and interference: Kuhl, 1998.

193 Brain differences in people with dyslexia: Eden and Zeffiro, 1998; Horwitz, Rumsey, and Donohue, 1998.

194 Exaggerated sounds for children with dyslexia: Merzenich et al., 1996; Tallal et al., 1996; Tallal et al., 1998.

194 Social influences on learning in birds: Eales, 1985; Baptista and Petrinovich, 1986.

Chapter Seven: Trailing Clouds of Glory

198 White House Conference on Early Learning and the Brain: "Studies show talking with infants shapes basis of ability to think," *The New York Times,* April 17, 1997; "Experts describe new research on early learning," *The Washington Post,* April 18, 1997.

198 What should we tell policy makers?: Many groups in the United States and other countries are grappling with the problem of translating the scientific discoveries to policy recommendations. In addition to the National Institutes of Health and the National Science Foundation, information is available from the Carnegie Foundation (P.O. Box 753, Waldorf, MD 20604), the Education Commission of the States (707 17th Street, Suite 2700, Denver, CO 80202), the Dana Alliance for Brain Initiatives (745 Fifth Ave., Suite 700, New York, NY 10151), the Parents as Teachers National

Center (10176 Corporate Square Drive, Suite 230, St. Louis, MO 63132), and Zero to Three: National Center for Infants, Toddlers and Families (734 15th Street, NW, Suite 1000, Washington, D.C. 20005).

203 Radical changes in the child-rearing environment due to changes in society: Skolnick and Skolnick, 1992; Hernandez and Myers, 1993; Scarr, 1998.

208 Do babies feel pain?: Barr, 1992, 1994; Wellington and Rieder, 1993.

209 "A time when meadow . . .": Wordsworth, 1943; "A world in a grain of sand . . .": Blake, 1965.

REFERENCES

Abramson, A. S., and L. Lisker. 1970. Discriminability along the voicing continuum: Cross-language tests. In *Proceedings of the Sixth International Congress of Phonetic Sciences, Prague 1967*, 569–73. Prague: Academia.

Ainsworth, M. D., M. C. Blehar, E. Waters, and S. Wall. 1978. *Patterns of attachment: A psychological study of the strange situation.* Hillsdale, N.J.: Erlbaum.

Ariès, P. 1962. *Centuries of childhood.* London: Jonathan Cape.

Aristotle. 1956. *De anima.* Trans. D. Ross. Oxford: Clarendon Press.

Aslin, R. N. 1987. Visual and auditory development in infancy. In *Handbook of infant development.* 2nd ed., ed. J. D. Osofsky, 5–97. New York: Wiley.

———. 1993. Segmentation of fluent speech into words: Learning models and the role of maternal input. In *Developmental neurocognition: Speech and face processing in the first year of life*, ed. B. de Boysson-Bardies, S. de Schonen, P. Jusczyk, P. McNeilage, and J. Morton, 305–16. Dordrecht, Netherlands: Kluwer.

Astington, J. W. 1993. *The child's discovery of the mind.* Cambridge: Harvard University Press.

Astington, J. W., P. L. Harris, and D. R. Olson. 1988. *Developing theories of mind.* New York: Cambridge University Press.

Astington, J. W., and J. Pelletier. 1996. The language of the mind: Its role in teaching and learning. In *Handbook of education and human development: New models of learning, teaching and schooling,* ed. D. R. Olson and N. Torrance, 593–619. Oxford: Blackwell.

Atkinson, J. 1984. Human visual development over the first 6 months of life: A review and a hypothesis. *Human Neurobiology* 3:61–74.

Atran, S. 1990. *Cognitive foundations of natural history: Towards an anthropology of science.* New York: Cambridge University Press.

Augustine, Saint. 1995. *Confessions.* Trans. G. Clark. Cambridge: Cambridge University Press.

Bahrick, L. E. 1987. Infants' intermodal perception of two levels of temporal structure in natural events. *Infant Behavior and Development* 10: 387–416.

Baillargeon, R. 1993. The object concept revisited: New directions in the investigation of infants' physical knowledge. In *Visual perception and cognition in infancy,* ed. C. Granrud, 265–315. Hillsdale, N.J.: Erlbaum.

Baillargeon, R., and M. Graber. 1987. Where's the rabbit? 5.5-month-old infants' representation of the height of a hidden object. *Cognitive Development* 2:375–92.

Baldwin, D. A. 1993a. Early referential understanding: Infants' ability to recognize referential acts for what they are. *Developmental Psychology* 29:832–43.

———. 1993b. Infants' ability to consult the speaker for clues to word reference. *Journal of Child Language* 20:395–418.

Baldwin, D. A., E. M. Markman, B. Bill, R. N. Desjardins, J. M. Irwin, and G. Tidball. 1996. Infants' reliance on a social criterion for establishing word-object relations. *Child Development* 67:3135–53.

Baldwin, D. A., E. M. Markman, and R. L. Melartin. 1993. Infants' ability to draw inferences about nonobvious object properties: Evidence from exploratory play. *Child Development* 64:711–28.

Baldwin, D. A., and L. J. Moses. 1994. Early understanding of referential intent and attentional focus: Evidence from language and emotion. In *Children's early understanding of mind: Origins and development,* ed. C. Lewis and P. Mitchell, 133–56. Hillsdale, N.J.: Erlbaum.

Bamberg, P. G., and M. A. Mandel. 1991. Adaptable phoneme-based models for large-vocabulary speech recognition. *Speech Communication* 10:437–51.

Banks, M. S. 1980. The development of visual accommodation during early infancy. *Child Development* 51:646–66.

Baptista, L. F., and L. Petrinovich. 1986. Song development in the white-crowned sparrow: Social factors and sex differences. *Animal Behaviour* 34:1359–71.

Baron-Cohen, S. 1995. *Mindblindness: An essay on autism and theory of mind.* Cambridge: MIT Press.

Baron-Cohen, S., H. Tager-Flusberg, and D. J. Cohen, eds. 1993. *Understanding other minds: Perspectives from autism.* New York: Oxford University Press.

———. 1999. *Understanding other minds: Perspectives from autism and developmental cognitive neuroscience.* Oxford: Oxford University Press.

Barr, R. G. 1992. Is this infant in pain? Caveats from the clinical setting. *American Pain Society* 1:187–90.

———. 1994. Pain experience in children: Developmental and clinical characteristics. In *Textbook of pain,* ed. P. D. Wall and R. Melzack, 739–65. New York: Churchill Livingstone.

Bartsch, K., and H. M. Wellman. 1995. *Children talk about the mind.* New York: Oxford University Press.

Bates, E., L. Benigni, I. Bretherton, L. Camaioni, and V. Volterra. 1979. *The emergence of symbols: Cognition-communication in infancy.* New York: Academic Press.

Bates, E., I. Bretherton, and L. Snyder. 1987. *From first words to grammar: Individual differences and dissociable mechanism.* New York: Cambridge University Press.

Bavelier, D., D. Corina, P. Jezzard, S. Padmanabhan, V. P. Clark, A. Karni, A. Prinster, A. Braun, A. Lalwani, J. P. Rauschecker, R. Turner, and H. Neville. 1997. Sentence reading: A functional MRI study at 4 Tesla. *Journal of Cognitive Neuroscience* 9:664–86.

Bellugi, U., P. P. Wang, and T. L. Jernigan. 1994. Williams syndrome: An unusual neuropsychological profile. In *Atypical cognitive deficits in developmental disorders: Implications for brain function,* ed. S. H. Broman and J. Grafman, 23–56. Hillsdale, N.J.: Erlbaum.

Bennett, P. M., and P. H. Harvey. 1985. Brain size, development and metabolism in birds and mammals. *Journal of Zoology* 207:491–509.

Berkeley, G. 1910. *An essay toward a new theory of vision.* Dublin: Pepyat.

Berko, J. 1958. The child's learning of English morphology. *Word* 14: 150–77.

Bertenthal, B. I. 1993. Perception of biomechanical motions: Intrinsic image and knowledge-based constraints. In *Carnegie Mellon symposia on cognition: Visual perception and cognition in infancy,* ed. C. Granrud, 175–214. Hillsdale, N.J.: Erlbaum.

Bertenthal, B. I., D. R. Proffitt, and S. J. Kramer. 1987. Perception of biomechanical motions by infants: Implementation of various processing constraints. *Journal of Experimental Psychology: Human Perception and Performance* 13 (Special issue: *The Ontogenesis of Perception*): 577–85.

Bertenthal, B. I., D. R. Proffitt, N. B. Spetner, and M. A. Thomas. 1985. The development of infant sensitivity to biomechanical motions. *Child Development* 56:531–43.

Bickerton, D. 1981. *The roots of language.* Ann Arbor, Mich.: Karoma.

———. 1990. *Language and species.* Chicago: University of Chicago Press.

———. 1995. *Language and human behavior.* Seattle: University of Washington Press.

Bjorklund, D. F. 1997. The role of immaturity in human development. *Psychological Bulletin* 122:153–69.

Blake, W. 1965. "Auguries of innocence." In *The poetry and prose of William Blake,* ed. D. Erdman. Garden City, N.Y.: Doubleday.

Block, N. 1990. The computer model of the mind. In *An invitation to cognitive science,* ed. D. N. Osherson and E. E. Smith. Vol. 3, *Thinking,* 247–89. Cambridge: MIT Press.

Bloom, L. 1973. *One word at a time: The use of single word utterances before syntax.* The Hague: Mouton.

———. 1991. *Language development from two to three.* New York: Cambridge University Press.

———. 1993. *The transition from infancy to language: Acquiring the power of expression.* New York: Cambridge University Press.

Blount, B. G., and E. J. Padgug. 1977. Prosodic, paralinguistic, and interactional features in parent-child speech: English and Spanish. *Journal of Child Language* 4:67–86.

Bomba, P. C., and E. R. Siqueland. 1983. The nature and structure of infant form categories. *Journal of Experimental Child Psychology* 35:294–328.

Bower, T. G. R. 1982. *Development in infancy.* 2nd ed. San Francisco: W. H. Freeman.

———. 1989. *The rational infant: Learning in infancy.* San Francisco: W. H. Freeman.

Bowerman, M. 1989. Learning a semantic system: What role do cognitive predispositions play? In *The teachability of language,* ed. M. Rice and R. L. Schiefelbusch, 133–69. Baltimore: Paul Brookes.

Bowlby, J. 1969. *Attachment and loss.* Vol. 1, *Attachment.* New York: Basic Books.

———. 1973. *Attachment and loss.* Vol. 2, *Separation: Anxiety and anger.* New York: Basic Books.

Boyer, P. 1994. *The naturalness of religious ideas: A cognitive theory of religion.* Berkeley: University of California Press.

Brainard, M. S., and E. I. Knudsen. 1998. Sensitive periods for visual calibration of the auditory space map in the barn owl optic tectum. *Journal of Neuroscience* 18:3929–42.

Bråten, S. 1999. *Intersubjective communication and emotion in early ontogeny.* Cambridge: Cambridge University Press.

Brazelton, T. B., B. Koslowski, and M. Main. 1974. The origins of reciprocity: The early mother-infant interaction. In *The effect of the infant on its caregiver,* ed. M. Lewis and L. A. Rosenblum, 49–76. New York: Wiley.

Brazelton, T. B., and E. Tronick. 1980. Preverbal communication between mothers and infants. In *The social foundations of language and thought,* ed. D. R. Olson, 299–315. New York: Norton.

Brentano, F. 1973. *Psychology from an empirical standpoint.* Trans. A. C. Rancurello, D. B. Terrell, and L. L. McAlister. London: Routledge & Kegan Paul.

Bretherton, I., and E. Waters. 1985. Growing points of attachment theory and research. *Monographs of the Society for Research in Child Development* 50, nos. 1–2 (serial no. 209).

Brothers, L. 1989. A biological perspective on empathy. *American Journal of Psychiatry* 146:10–19.

Brown, J. R., N. Donelan-McCall, and J. Dunn. 1996. Why talk about mental states? The significance of children's conversations with friends, siblings, and mothers. *Child Development* 67:836–49.

Brown, R. 1973. *A first language: The early stages.* Cambridge: Harvard University Press.

Bruner, J. S. 1972. Nature and uses of immaturity. *American Psychologist* 27:1–23.

———. 1973. Organization of early skilled action. *Child Development* 44: 1–11.

———. 1975. From communication to language—A psychological perspective. *Cognition* 3:255–87.

———. 1983. *Child's talk: Learning to use language.* New York: Norton.

———. 1990. *Acts of meaning.* Cambridge: Harvard University Press.

———. 1996. *The culture of education.* Cambridge: Harvard University Press.

Bryant, P. E., P. Jones, V. Claxton, and G. M. Perkins. 1972. Recognition of shapes across modalities by infants. *Nature* 240:303–4.

Bullock, M., and R. Gelman. 1979. Preschool children's assumptions about cause and effect: Temporal ordering. *Child Development* 50:89–96.

Bushnell, I. W. R., F. Sai, and J. T. Mullin. 1989. Neonatal recognition of the mother's face. *British Journal of Developmental Psychology* 7:3–15.

Butterworth, G. 1997. Starting point: Finger pointing by babies is correlated with the rate of language acquisition. *Natural History* 106:14–16.

Butterworth, G., and N. Jarrett. 1982. Piaget's stage 4 error: Background to the problem. *British Journal of Psychology* 73:175–85.

Byrne, R. W., and A. Whiten, eds. 1988. *Machiavellian intelligence: Social expertise and the evolution of intellect in monkeys, apes and humans.* Oxford: Clarendon Press.

Campos, J. J., and C. R. Stenberg. 1981. Perception, appraisal and emotion: The onset of social referencing. In *Infant social cognition: Empirical and theoretical considerations,* ed. M. E. Lamb and L. R. Sherrod, 273–314. Hillsdale, N.J.: Erlbaum.

Caplan, D. 1992. *Language: Structure, processing, and disorders.* Cambridge: MIT Press.

Carey, S. 1978. The child as a word learner. In *Linguistic theory and psychological reality,* ed. M. Halle, J. Bresnan, and G. A. Miller, 264–93. Cambridge: MIT Press.

———. 1985. *Conceptual change in childhood.* Cambridge: MIT Press.

———. 1988. Conceptual differences between children and adults. *Mind and Language* 3:167–81.

Carpenter, M., K. Nagell, and M. Tomasello. 1998. Social cognition, joint attention, and communicative competence from 9 to 15 months of age. *Monographs of the Society for Research in Child Development* 63, no. 4 (serial no. 255).

Carroll, J. B. 1956. *Language, thought, and reality: Selected writings of Benjamin Lee Whorf.* Cambridge: MIT Press.

Carruthers, P. 1996. *Language, thought, and consciousness: An essay in philosophical psychology.* Cambridge: Cambridge University Press.

Chalmers, D. 1996. *The conscious mind: In search of a fundamental theory.* New York: Oxford University Press.

Chandler, M., A. S. Fritz, and S. Hala. 1989. Small scale deceit: Deception as a marker of two-, three-, and four-year-olds' early theories of mind. *Child Development* 60:1263–77.

Chapman, R. S., N. W. Streim, E. R. Crais, D. Salmon, E. A. Strand, and N. A. Negri. 1992. Child talk: Assumptions of a developmental process model for early language learning. In *Processes in language acquisition and disorders,* ed. R. S. Chapman, 3–19. St. Louis: Mosby Year Book.

Cheour, M., R. Ceponiene, A. Lehtokoski, A. Luuk, J. Allik, K. Alho, and R. Näätänen. 1998. Development of language-specific phoneme representations in the infant brain. *Nature Neuroscience* 1:351–53.

Choi, S., and A. Gopnik, 1995. Early acquisition of verbs in Korean: A cross-linguistic study. *Journal of Child Language* 22:497–529.

Chomsky, N. 1957. *Syntactic structures.* The Hague: Mouton.

———. 1965. *Aspects of the theory of syntax.* Cambridge: MIT Press.

———. 1980. *Rules and representations.* New York: Columbia University Press.

Chugani, H. T. 1994. Development of regional brain glucose metabolism in relation to behavior and plasticity. In *Human behavior and the developing brain,* ed. G. Dawson and K. W. Fischer, 153–75. New York: Guilford.

———. 1998. A critical period of brain development: Studies of cerebral glucose utilization with PET. *Preventive Medicine* 27:184–88.

Chugani, H. T., M. E. Phelps, and J. C. Mazziotta. 1987. Positron emission tomography study of human brain functional development. *Annals of Neurology* 22:487–97.

Churchland, P. M. 1981. Eliminative materialism and the propositional attitudes. *Journal of Philosophy* 78:67–90.

———. 1988. *Matter and consciousness: A contemporary introduction to the philosophy of mind.* Rev. ed. Cambridge: MIT Press.

———. 1995. *The engine of reason, the seat of the soul: A philosophical journey into the brain.* Cambridge: MIT Press.

Churchland, P. S. 1986. *Neurophilosophy: Toward a unified science of the mind-brain.* Cambridge: MIT Press.

Clark, A. 1993. *Associative engines: Connectionism, concepts, and representational change.* Cambridge: MIT Press.

Clark, E. V. 1974. Some aspects of the conceptual basis for first language acquisition. In *Language perspectives—Acquisition, retardation and intervention,* ed. R. L. Schiefelbusch and L. L. Lloyd, 105–28. Baltimore: University Park Press.

Cohen, L. B. 1979. Our developing knowledge of infant perception and cognition. *American Psychologist* 34:894–99.

Cole, M. 1996. *Cultural psychology: A once and future discipline.* Cambridge: Harvard University Press.

Crystal, D. 1997. *The Cambridge encyclopedia of language.* Cambridge: Cambridge University Press.

Curtiss, S. 1977. *Genie: A psycholinguistic study of a modern day "wild child."* New York: Academic.

Cutler, A., and S. Butterfield. 1992. Rhythmic cues to speech segmentation: Evidence from juncture misperception. *Journal of Memory and Language* 31:218–36.

Damasio, A. R., and H. Damasio. 1992. Brain and language. *Scientific American* 267:88–95.

D'Andrade, R. G. 1995. *The development of cognitive anthropology.* Cambridge: Cambridge University Press.

Davidson, D. 1980. *Essays on action and events.* New York: Oxford University Press.

Dawson, G., A. N. Meltzoff, J. Osterling, and J. Rinaldi. 1998. Neurophysiological correlates of early symptoms of autism. *Child Development* 69:1276–85.

de Boysson-Bardies, B. 1993. Ontogeny of language-specific syllabic productions. In *Developmental neurocognition: Speech and face processing in the first year of life,* ed. B. de Boysson-Bardies, S. de Schonen, P. Jusczyk, P. McNeilage, and J. Morton, 353–63. Dordrecht, Netherlands: Kluwer.

Deacon, T. W. 1997. *The symbolic species: The co-evolution of language and the brain.* New York: Norton.

DeCasper, A. J., and W. P. Fifer. 1980. Of human bonding: Newborns prefer their mothers' voices. *Science* 208:1174–76.

Dennett, D. C. 1978. *Brainstorms: Philosophical essays on mind and language.* Cambridge: MIT Press.

———. 1991. *Consciousness explained.* Boston: Little, Brown & Co.

Dennis, M., and H. A. Whitaker. 1976. Language acquisition following hemidecortication: Linguistic superiority of the left over the right hemisphere. *Brain and Language* 3:404–33.

Descartes, R. 1911. *The philosophical works of Descartes.* Trans. E. S. Haldane and G. R. T. Ross. Vol. 1. Cambridge: Cambridge University Press.

———. 1952. *Descartes' philosophical writings.* Trans. and ed. N. Kemp Smith. London: Macmillan.

Desimone, R., T. D. Albright, C. G. Gross, and C. Bruce. 1984. Stimulus-selective properties of inferior temporal neurons in the macaque. *Journal of Neuroscience* 8:2051–62.

Desrochers, S., P. Morissette, and M. Ricard. 1995. Two perspectives on pointing in infancy. In *Joint attention: Its origins and role in development,* ed. C. Moore and P. Dunham, 85–101. Hillsdale, N.J.: Erlbaum.

Diamond, M. C., D. Krech, and M. R. Rosenzweig. 1964. The effects of an enriched environment on the histology of the rat cerebral cortex. *Journal of Comparative Neurology* 123:111–20.

Doupe, A., and P. K. Kuhl. 1999. Birdsong and human speech: Common themes and mechanisms. *Annual Review of Neuroscience* 22:567–631.

Dreyfus, H. 1992. *What computers still can't do: A critique of artificial reason.* Cambridge: MIT Press.

Duchowny, M., P. Jayakar, A. S. Harvey, T. Resnick, L. Alvarez, P. Dean, and B. Levin. 1996. Language cortex representation: Effects of developmental versus acquired pathology. *Annals of Neurology* 40:31–38.

Dunn, J., and R. Plomin. 1990. *Separate lives: Why siblings are so different.* New York: Basic Books.

Eales, L. A. 1985. Song learning in zebra finches: Some effects of song model availability on what is learnt and when. *Animal Behavior* 37: 507–8.

Edelman, G. M. 1992. *Bright air, brilliant fire: On the matter of the mind.* New York: Basic Books.

Eden, G. F., and T. A. Zeffiro. 1998. Neural systems affected in developmental dyslexia revealed by functional neuroimaging. *Neuron* 21: 279–82.

Eimas, P. D. 1975. Auditory and phonetic coding of the cues for speech: Discrimination of the /r-l/ distinction by young infants. *Perception and Psychophysics* 18:341–47.

Eimas, P. D., E. R. Siqueland, P. Jusczyk, and J. Vigorito. 1971. Speech perception in infants. *Science* 171:303–6.

Elman, J. L., E. A. Bates, M. H. Johnson, A. Karmiloff-Smith, D. Parisi, and K. Plunkett, eds. 1996. *Rethinking innateness: A connectionist perspective on development.* Cambridge: MIT Press.

Eriksson, P. S., E. Perfilieva, T. Bjork-Eriksson, A. M. Alborn, C. Nordborg, D. A. Peterson, and F. H. Gage. 1998. Neurogenesis in the adult human hippocampus. *Nature Medicine* 4:1313–17.

Estes, W. K. 1993. Concepts, categories, and psychological science. *Psychological Science* 4:143–53.

Eulitz, C., B. Maess, C. Pantev, and A. D. Friederici. 1996. Oscillatory neuromagnetic activity induced by language and non-language stimuli. *Cognitive Brain Research* 4:121–32.

Evans-Pritchard, E. E. 1976. *Witchcraft, oracles, and magic among the Azande.* Oxford: Clarendon.

Fantz, R. L. 1963. Pattern vision in newborn infants. *Science* 140:296–97.

Fenson, L., P. S. Dale, J. S. Reznick, E. Bates, D. Thal, and S. J. Pethick. 1994. Variability in early communicative development. *Monographs of the Society for Research in Child Development* 59, no. 5 (serial no. 242).

Ferguson, C. A. 1964. Baby talk in six languages. *American Anthropologist* 66:103–14.

Ferguson, C. A., L. Menn, and C. Stoel-Gammon, eds. 1992. *Phonological development: Models, research, implications.* Timonium, Md.: York Press.

Fernald, A. 1985. Four-month-old infants prefer to listen to motherese. *Infant Behavior and Development* 8:181–95.

———. 1992. Human maternal vocalizations to infants as biologically relevant signals: An evolutionary perspective. In *The adapted mind: Evolutionary psychology and the generation of culture,* ed. J. H. Barkow, L. Cosmides, and J. Tooby, 391–428. New York: Oxford University Press.

Fernald, A., and P. Kuhl. 1987. Acoustic determinants of infant preference for motherese speech. *Infant Behavior and Development* 10:279–93.

Fernald, A., and T. Simon. 1984. Expanded intonation contours in mothers' speech to newborns. *Developmental Psychology* 20:104–13.

Fernald, A., T. Taeschner, J. Dunn, M. Papousek, B. de Boysson-Bardies, and I. Fukui. 1989. A cross-language study of prosodic modification in mothers' and fathers' speech to preverbal infants. *Journal of Child Language* 16:477–501.

Feyerabend, P. K. 1975. *Against method.* New York: Verso.

Field, T. M., D. Cohen, R. Garcia, and R. Greenberg. 1984. Mother-stranger face discrimination by the newborn. *Infant Behavior and Development* 7:19–25.

Flavell, J. H. 1963. *The developmental psychology of Jean Piaget.* New York: Van Nostrand.

———. 1999. Cognitive development: Children's knowledge about the mind. *Annual Review of Psychology* 50, 21–45.

Flavell, J. H., B. A. Everett, K. Croft, and E. R. Flavell. 1981. Young children's knowledge about visual perception: Further evidence for the Level 1–Level 2 distinction. *Developmental Psychology* 17:99–103.

Flavell, J. H., F. L. Green, and E. R. Flavell. 1986. Development of knowledge about the appearance–reality distinction. *Monographs of the Society for Research in Child Development* 51, no. 1 (serial no. 212).

Flavell, J. H., and P. H. Miller. 1998. Social cognition. In *Handbook of child psychology*, ed. W. Damon. Vol. 2, *Cognition, perception, and language*, ed. D. Kuhn and R. Siegler, 851–98. New York: Wiley.

Flege, J. E. 1988. Factors affecting degree of perceived foreign accent in English sentences. *Journal of the Acoustical Society of America* 84:70–79.

Fletcher, P., and B. MacWhinney, eds. 1995. *The handbook of child language*. Cambridge, Mass.: Blackwell.

Fodor, J. A. 1975. *The language of thought*. New York: Thomas Y. Crowell.

Ford, K. M., C. N. Glymour, and P. T. Hayes, eds. 1995. *Android epistemology*. Cambridge: MIT Press.

Forguson, L., and A. Gopnik. 1988. The ontogeny of common sense. In *Developing theories of mind*, ed. J. W. Astington, P. L. Harris, and D. R. Olson, 226–43. New York: Cambridge University Press.

Franco, F., and G. Butterworth. 1996. Pointing and social awareness: Declaring and requesting in the second year. *Journal of Child Language* 23:307–36.

Frege, G. 1952. On sense and meaning. In *Translations from philosophical writings of Gottlob Frege*, ed. P. Geach and M. Black, 56–78. Oxford: Basil Blackwell.

Freud, S. 1953. Three essays on the theory of sexuality. In *The standard edition of the complete psychological works of Sigmund Freud*, trans J. Strachey, vol. 7, 123–245. London: Hogarth Press.

Frey, K. A., S. Minoshima, and D. E. Kuhl. 1998. Neurochemical imaging of Alzheimer's disease and other degenerative dementias. *Quarterly Journal of Nuclear Medicine* 42:166–68.

Fromkin, V., S. Krashen, S. Curtis, D. Rigler, and M. Rigler. 1974. The development of language in Genie: A case of language acquisition beyond the "critical period." *Brain and Language* 1:81–107.

Fullard, W., and A. M. Reiling. 1976. An investigation of Lorenz's "babyness." *Child Development* 147:1191–93.

Gallen, C. C., B. J. Schwartz, R. D. Bucholz, G. Malik, G. L. Barkley, J. Smith, H. Tung, B. Copeland, L. Bruno, and S. Assam. 1995. Presurgical localization of functional cortex using magnetic source imaging. *Journal of Neurosurgery* 82:988–94.

Ganslandt, O., R. Steinmeier, H. Kober, J. Vieth, J. Kassubek, J. Romstock, C. Strauss, and R. Fahlbusch. 1997. Magnetic source imaging combined with image-guided frameless stereotaxy: A new method in surgery around the motor strip. *Neurosurgery* 41:621–27.

Gardner, H. 1985. *The mind's new science: A history of the cognitive revolution.* New York: Basic Books.

———. 1991. *The unschooled mind: How children think and how schools should teach.* New York: Basic Books.

———. 1995. *Leading minds: An anatomy of leadership.* New York: Basic Books.

———. 1997. *Extraordinary minds: Portraits of exceptional individuals and an examination of our extraordinariness.* New York: Basic Books.

Gelman, R., M. Bullock, and E. Meck. 1980. Preschoolers' understanding of simple object transformations. *Child Development* 51:691–99.

Gelman, R., and E. M. Williams. 1998. Enabling constraints for cognitive development and learning: Domain specificity and epigenesis. In *Handbook of child psychology,* ed. W. Damon. Vol. 2, *Cognition, perception, and language,* ed. D. Kuhn and R. Siegler, 575–630. New York: Wiley.

Gelman, S. A., and J. D. Coley. 1991. Language and categorization: The acquisition of natural kind terms. In *Perspectives on language and thought: Interrelations in development,* ed. S. A. Gelman and J. P. Byrnes, 146–96. New York: Cambridge University Press.

Gelman, S. A., and E. M. Markman. 1987. Young children's inductions from natural kinds: The role of categories and appearances. *Child Development* 58:1532–41.

Gelman, S. A., and T. Tardif. 1998. Acquisition of nouns and verbs in Mandarin and English. In *The proceedings of the twenty-ninth annual child language forum,* ed. E. V. Clark, 27–36. Stanford, Calif.: Center for the Study of Language and Information.

Gelman, S. A., and H. M. Wellman. 1991. Insides and essence: Early understandings of the non-obvious. *Cognition* 38:213–44.

Gergely, G., Z. Nádasdy, G. Csibra, and S. Bíró. 1995. Taking the intentional stance at 12 months of age. *Cognition* 56:165–93.

Geschwind, N. 1979. Specializations of the human brain. *Scientific American* 241:180–99.

Geschwind, N., and A. Galaburda. 1987. *Cerebral lateralization: Biological mechanism, associations, and pathology.* Cambridge: MIT Press.

Gibson, E. J. 1987. What does infant perception tell us about theories of perception? *Journal of Experimental Psychology: Human Perception and Performance* 13:515–23.

Gibson, E. J., and A. S. Walker. 1984. Development of knowledge of visual-tactual affordances of substance. *Child Development* 55:453–60.

Glymour, C., J. Ramsey, and T. Roush. 1999. *Automated mineral classification from near infrared reflectance spectra.* Technical report, Department of Philosophy, Carnegie Mellon University, Pittsburgh.

Goldman, D., and D. Homa. 1977. Integrative and metric properties of abstracted information as a function of category discriminability, instance variability, and experience. *Journal of Experimental Psychology: Human Learning and Memory* 3:375–85.

Golinkoff, R., C. B. Mervis, and K. Hirsh-Pasek. 1994. Early object labels: The case for a developmental lexical principles framework. *Journal of Child Language* 21:125–55.

Goodglass, H. 1993. *Understanding aphasia.* San Diego: Academic Press.

Goodman, N. 1983. *Fact, fiction, and forecast.* 4th ed. Cambridge: Harvard University Press.

Gopnik, A. 1982. Words and plans: Early language and the development of intelligent action. *Journal of Child Language* 9:303–18.

———. 1984. The acquisition of *gone* and the development of the object concept. *Journal of Child Language* 11:273–92.

———. 1988a. Conceptual and semantic development as theory change: The case of object permanence. *Mind and Language* 3:197–216.

———. 1988b. Three types of early words: The emergence of social words, names and cognitive-relational words in the one-word stage and their relation to cognitive development. *First Language* 8:49–69.

———. 1993. How we know our minds: The illusion of first-person knowledge of intentionality. *Behavioral and Brain Sciences* 16:1–14.

———. 1996a. The post-Piaget era. *Psychological Science* 7:221–25.

———. 1996b. The scientist as child. *Philosophy of Science* 63:485–514.

———. 1998. Explanation as orgasm. *Minds and Machines* 8:101–18.

Gopnik, A., and J. W. Astington. 1988. Children's understanding of representational change and its relation to the understanding of false belief and appearance-reality distinction. *Child Development* 59:26–37.

Gopnik, A., and S. Choi, 1990. Do linguistic differences lead to cognitive differences?: A cross-linguistic study of semantic and cognitive development. *First Language* 10:199–215.

Gopnik, A., S. Choi, and T. Baumberger. 1996. Cross-linguistic differences in early semantic and cognitive development. *Cognitive Development* 11:197–227.

Gopnik, A., J. Esterly, and A. N. Meltzoff. 1995. Very young children's understanding of visual perspective taking. First annual west coast conference on theory of mind, Feb. 3–4, Eugene, Oreg.

Gopnik, A., and P. Graf. 1988. Knowing how you know: Young children's ability to identify and remember the sources of their beliefs. *Child Development* 59:1366–71.

Gopnik, A., and A. N. Meltzoff. 1984. Semantic and cognitive development in 15- to 21-month-old children. *Journal of Child Language* 11: 495–513.

———. 1986. Relations between semantic and cognitive development in the one-word stage: The specificity hypothesis. *Child Development* 57:1040–53.

———. 1987. The development of categorization in the second year and its relation to other cognitive and linguistic developments. *Child Development* 58:1523–31.

———. 1992. Categorization and naming: Basic-level sorting in eighteen-month-olds and its relation to language. *Child Development* 63:1091–1103.

———. 1994. Minds, bodies, and persons: Young children's understanding of the self and others as reflected in imitation and theory of mind research. In *Self-awareness in animals and humans: Developmental perspectives,* ed. S. T. Parker, R. W. Mitchell, and M. L. Boccia, 166–86. New York: Cambridge University Press.

———. 1997. *Words, thoughts, and theories.* Cambridge: MIT Press.

Gopnik, A., and V. Slaughter. 1991. Young children's understanding of changes in their mental states. *Child Development* 62:98–110.

Gopnik, A., V. Slaughter, and A. N. Meltzoff. 1994. Changing your views: How understanding visual perception can lead to a new theory of the mind. In *Children's early understanding of mind: Origins and development,* ed. C. Lewis and P. Mitchell, 157–81. Hillsdale, N.J.: Erlbaum.

Gopnik, A., and D. Sobel. 1997. Detecting blickets. Poster presented at the meeting of the Society for Research in Child Development, April 3–7, Washington, D.C.

Gopnik, A., and H. M. Wellman. 1994. The theory theory. In *Mapping the mind: Domain specificity in cognition and culture,* ed. L. A. Hirschfeld and S. A. Gelman, 257–93. New York: Cambridge University Press.

Gopnik, M. 1990. Dysphagia in an extended family. *Nature* 344:715.

Gopnik, M., and M. Crago. 1991. Familial aggregation of a developmental language disorder. *Cognition* 39:1–50.

Goto, H. 1971. Auditory perception by normal Japanese adults of the sounds "l" and "r." *Neuropsychologia* 9:317–23.

Gottlieb, G., and N. Krasnegor. 1985. *Measurement of audition and vision in the first year of postnatal life: A methodological overview.* Norwood, N.J.: Ablex.

Gould, E., P. Tanapat, B. S. McEwen, G. Flugge, and E. Fuchs. 1998. Proliferation of granule cell precursors in the dentate gyrus of adult monkeys is diminished by stress. *Proceedings of the National Academy of Sciences of the United States of America* 95:3168–71.

Gould, S. J. 1977. *Ontogeny and phylogeny.* Cambridge: Harvard University Press.

Grandin, T. 1995. *Thinking in pictures: and other reports from my life with autism.* New York: Vintage Books.

Granrud, C. E. 1987. Size constancy in newborn human infants. *Investigative Ophthalmology and Visual Science,* supplement 28:5.

Greenough, W. T., F. Volkman, and J. M. Juraska. 1973. Effects of rearing complexity on dendritic branching in frontolateral and temporal cortex of the rat. *Experimental Neurology* 41:371–78.

Grieser, D. L., and P. K. Kuhl. 1988. Maternal speech to infants in a tonal language: Support for universal prosodic features in motherese. *Developmental Psychology* 24:14–20.

Habermas, J. 1979. *Communication and the evolution of society.* Trans. T. McCarthy. Boston: Beacon Press.

Haith, M. M. 1980. *Rules that babies look by: The organization of newborn visual activity.* Hillsdale, N.J.: Erlbaum.

———. 1998. Who put the cog in infant cognition? Is rich interpretation too costly? *Infant Behavior and Development* 21:167–79.

Hanna, E., and A. N. Meltzoff. 1993. Peer imitation by toddlers in laboratory, home, and day-care contexts: Implications for social learning and memory. *Developmental Psychology* 29:701–10.

Happé, F. 1995. *Autism: An introduction to psychological theory.* Cambridge: Harvard University Press.

Haralick, R. M., and L. G. Shapiro. 1992. *Computer and robot vision.* 2 vols. Reading, Mass.: Addison-Wesley.

Harris, P. L. 1987. The development of search. In *Handbook of infant perception,* ed. P. Salapatek and L. Cohen. Vol. 2, *From perception to cognition,* 155–207. New York: Academic Press.

———. 1989. *Children and emotion: The development of psychological understanding.* Oxford: Basil Blackwell.

Haugeland, J., ed. 1981. *Mind design: Philosophy, psychology, and artificial intelligence.* Cambridge: MIT Press.

———. 1989. *Artificial intelligence: The very idea.* Cambridge: MIT Press.

———. 1997. *Mind design II: Philosophical, psychological, artificial intelligence.* Cambridge: MIT Press.

Hauser, M. D. 1996. *The evolution of communication.* Cambridge: MIT Press.

Hempel, C. G. 1965. *Aspects of scientific explanation and other essays in the philosophy of science.* New York: Free Press.

Herken, R., ed. 1988. *The universal Turing machine: A half-century survey.* Oxford: Oxford University Press.

Hernandez, D. J., and D. E. Myers. 1993. *America's children: Resources from family, government, and the economy.* New York: Russell Sage Foundation.

Hickling, A. K., and S. A. Gelman. 1995. How does your garden grow? Early conceptualization of seeds and their place in the plant growth cycle. *Child Development* 66:856–76.

Hildreth, E. C., and S. Ullman. 1989. The computational study of vision. In *Foundations of cognitive science,* ed. M. I. Posner, 581–630. Cambridge: MIT Press.

Hirsh-Pasek, K., and R. M. Golinkoff. 1996. *The origins of grammar: Evidence from early language comprehension.* Cambridge: MIT Press.

Hobbes, T. 1962. *Leviathan.* New York: Macmillan.

Hobson, R. P. 1993. *Autism and the development of mind.* Hillsdale, N.J.: Erlbaum.

Hofsten, C. V., and E. S. Spelke. 1985. Object perception and object-directed reaching in infancy. *Journal of Experimental Psychology: General* 114:198–212.

Hood, B. M., L. Murray, F. King, R. Hooper, J. Atkinson, and O. Braddick. 1996. Habituation changes in early infancy: Longitudinal measures from birth to 6 months. *Journal of Reproductive and Infancy Psychology* 14:177–85.

Horwitz, B., J. M. Rumsey, and B. C. Donohue. 1998. Functional connectivity of the angular gyrus in normal reading and dyslexia. *Proceedings of the National Academy of Sciences of the United States of America* 95:8939–44.

Hubel, D. H., and T. N. Wiesel. 1963. Receptive fields of cells in striate cortex of very young, visually inexperienced kittens. *Journal of Neurophysiology* 26:994–1002.

———. 1965. Binocular interaction in striate cortex of kittens reared with artificial squint. *Journal of Neurophysiology* 28:1041–59.

———. 1970. The period of susceptibility to the physiological effects of unilateral eye closure in kittens. *Journal of Physiology* 206:419–36.

Hume, D. 1984. *A treatise of human nature.* New York: Penguin.

Huttenlocher, P. R. 1979. Synaptic density in human frontal cortex: Developmental changes and effects of aging. *Brain Research* 163: 195–205.

———. 1990. Morphometric study of human cerebral cortex development. *Neuropsychologia* 28:517–27.

———. 1994. Synaptogenesis in human cerebral cortex. In *Human behavior and the developing brain,* ed. G. Dawson and K. W. Fischer, 35–54. New York: Guilford.

Huttenlocher, P. R., and C. de Courten. 1987. The development of synapses in striate cortex of man. *Human Neurobiology* 6:1–9.

Itard, J. M. 1962. *The wild boy of Aveyron.* Englewood Cliffs, N.J.: Prentice-Hall.

Iverson, P., and P. K. Kuhl. 1995. Mapping the perceptual magnet effect for speech using signal detection theory and multidimensional scaling. *Journal of the Acoustical Society of America* 97:553–62.

———. 1996. Influences of phonetic identification and category goodness on American listeners' perception of /r/ and /l/. *Journal of the Acoustical Society of America* 99:1130–40.

Jacobs, B., M. Schall, and A. B. Sheibel. 1993. A quantitative dendritic analysis of Wernicke's area in humans. II. Gender, hemispheric, and environmental factor. *Journal of Comparative Neurology* 327:97–111.

Jacobson, J. L., D. C. Boersma, R. B. Fields, and K. L. Olson. 1983. Paralinguistic features of adult speech to infants and small children. *Child Development* 54:436–42.

Jakobson, R. 1960. Why "mama" and "papa"? In *Perspectives in psychological theory,* ed. B. Kaplan and S. Wapner. 124–34. New York: International Universities Press.

Jenkins, J. M., and J. W. Astington. 1996. Cognitive factors and family structure associated with theory of mind development in young children. *Developmental Psychology* 32:70–78.

Johnson, J. S., and E. L. Newport. 1991. Critical period effects on universal properties of language: The status of subjacency in the acquisition of a second language. *Cognition* 39:215–58.

Johnson, S. C., and S. Carey. 1998. Knowledge enrichment and conceptual change in folkbiology: Evidence from Williams syndrome. *Cognitive Psychology* 37:156–200.

Jones, T. A., A. Y. Klintsova, V. L. Kilman, A. M. Sirevaag, and W. T. Greenough. 1997. Induction of multiple synapses by experience in the visual cortex of adult rats. *Neurobiology of Learning and Memory* 68: 13–20.

Jusczyk, P. W. 1985. The high-amplitude sucking technique as a methodological tool in speech perception research. In *Measurement of audition and vision in the first year of postnatal life: A methodological overview*, ed. G. Gottlieb and N. A. Krasnegor, 195–222. Norwood, N.J.: Ablex.

———. 1997. *The discovery of spoken language.* Cambridge: MIT Press.

Jusczyk, P. W., A. Cutler, and N. J. Redanz. 1993. Infants' preference for the predominant stress patterns of English words. *Child Development* 64:675–87.

Jusczyk, P. W., A. D. Friederici, J. M. I. Wessels, V. Y. Svenkerud, and A. M. Jusczyk. 1993. Infants' sensitivity to the sound patterns of native language words. *Journal of Memory and Language* 32:402–420.

Kagan, J. 1970. The determinants of attention in the infant. *American Scientist* 58:298–306.

———. 1998. *Three seductive ideas.* Cambridge: Harvard University Press.

Kalish, C. 1988. Reasons and causes: Children's understanding of conformity to social rules and physical laws. *Child Development* 69:706–20.

Karmiloff-Smith, A. 1992. *Beyond modularity: A developmental perspective on cognitive science.* Cambridge: MIT Press.

Karmiloff-Smith, A., and B. Inhelder. 1974. If you want to get ahead, get a theory. *Cognition* 3:195–212.

Kaye, K. L., and T. G. R. Bower. 1994. Learning and intermodal transfer of information in newborns. *Psychological Science* 5:286–88.

Keil, F. C. 1989. *Concepts, kinds, and cognitive development.* Cambridge: MIT Press.

Keil, F. C., and R. Wilson. 1998. Cognition and explanation *Minds and Machines* 8, no. 1 (special issue).

Kellman, P. J., and M. E. Arterberry. 1998. *The cradle of knowledge: Development of perception in infancy.* Cambridge: MIT Press.

Kellman, P. J., and M. S. Banks. 1998. Infant visual perception. In *Handbook of child psychology*, ed. W. Damon. Vol. 2, *Cognition, perception, and language*, ed. D. Kuhn and R. Siegler, 103–46. New York: Wiley.

Kellman, P. J., and E. S. Spelke. 1983. Perception of partly occluded objects in infancy. *Cognitive Psychology* 15:483–524.

Kent, R. D. 1992. The biology of phonological development. In *Phonological development: Models, research, implications*, ed. C. A. Ferguson, L. Menn, and C. Stoel-Gammon, 65–90. Timonium, Md.: York.

Kim, K. H. S., N. R. Relkin, K. M. Lee, and J. Hirsch. 1997. Distinct cortical areas associated with native and second languages. *Nature* 388:172–74.

Kirn, J. R., and F. Nottebohm. 1993. Direct evidence for loss and replacement of projection neurons in adult canary brain. *Journal of Neuroscience* 132:1654–63.

Kitcher, P. 1993. *Advancement of science: Science without legend, objectivity without illusions.* Oxford: Oxford University Press.

Klaus, M. H. 1998. Mother and infant: Early emotional ties. In *New perspectives in early emotional development,* ed. J. G. Warhol, 51–57. Skillman, N. J.: Johnson & Johnson Pediatric Institute.

Klaus, M. H., and J. H. Kennell. 1982. *Maternal-infant bonding.* St. Louis: Mosby.

Klein, P. J., and A. N. Meltzoff. 1999. Long-term memory, forgetting, and deferred imitation in 12-month-old infants. *Developmental Science* 2:102–13.

Knudsen, E. I. 1998. Capacity for plasticity in the adult owl auditory system expanded by juvenile experience. *Science* 279:1531–33.

Konishi, M. 1985. Birdsong: From behavior to neuron. *Annual Review of Neuroscience* 8:125–70.

Kosslyn, S., and R. A. Andersen. 1992. *Frontiers in cognitive neuroscience.* Cambridge: MIT Press.

Kripke, S. A. 1972. Naming and necessity. In *Semantics of natural language,* ed. D. Davidson and G. Harman, 253–355. Dordrecht, Netherlands: Reidel.

———. 1980. *Naming and necessity.* Cambridge: Harvard University Press.

Kuhl, P. K. 1985a. Categorization of speech by infants. In *Neonate cognition: Beyond the blooming buzzing confusion,* ed. J. Mehler and R. Fox, 231–62. Hillsdale, N.J.: Erlbaum.

———. 1985b. Methods in the study of infant speech perception. In *Measurement of audition and vision in the first year of postnatal life: A methodological overview,* ed. G. Gottlieb and N. Krasnegor, 223–51. Norwood, N.J.: Ablex.

———. 1987. Perception of speech and sound in early infancy. In *Handbook of infant perception,* ed. P. Salapatek and L. Cohen. Vol. 2, *From perception to cognition,* 275–382. New York: Academic Press.

———. 1988. Auditory perception and the evolution of speech. *Human Evolution* 3:19–43.

———. 1989. On babies, birds, modules, and mechanisms: A comparative approach to the acquisition of vocal communication. In *The comparative psychology of audition: Perceiving complex sounds,* ed. R. J. Dooling and S. H. Hulse, 379–419. Hillsdale, N.J.: Erlbaum.

———. 1991. Human adults and human infants show a "perceptual magnet effect" for the prototypes of speech categories, monkeys do not. *Perception and Psychophysics* 50:93–107.

———. 1994a. Speech perception. In *Introduction to communication sciences and disorders,* ed. F. D. Minifie, 77–148. San Diego: Singular.

———. 1994b. Learning and representation in speech and language. *Current Opinion in Neurobiology* 4:812–22.

———. 1998. The development of speech and language. In *Mechanistic relationships between development and learning,* ed. T. J. Carew, R. Menzel, and C. J. Shatz, 53–73. New York: Wiley.

Kuhl, P. K., J. E. Andruski, I. A. Chistovich, L. A. Chistovich, E. V. Kozhevnikova, V. L. Ryskina, E. I. Stolyarova, U. Sundberg, and F. Lacerda. 1997. Cross-language analysis of phonetic units in language addressed to infants. *Science* 277:684–86.

Kuhl, P. K., and P. Iverson. 1995. Linguistic experience and the "perceptual magnet effect." In *Speech perception and linguistic experience: Issues in cross-language research,* ed. W. Strange, 121–54. Timonium, Md.: York Press.

Kuhl, P. K., and A. N. Meltzoff. 1982. The bimodal perception of speech in infancy. *Science* 218:1138–41.

———. 1984. The intermodal representation of speech in infants. *Infant Behavior and Development* 7:361–81.

———. 1996. Infant vocalizations in response to speech: Vocal imitation and developmental change. *Journal of the Acoustical Society of America* 100:2425–38.

Kuhl, P. K., K. A. Williams, F. Lacerda, K. N. Stevens, and B. Lindblom. 1992. Linguistic experience alters phonetic perception in infants by 6 months of age. *Science* 255:606–8.

Kuhn, T. S. 1962. *The structure of scientific revolutions.* Chicago: University of Chicago Press.

Kurzweil, R. 1999. *The age of spiritual machines: When computers exceed human intelligence.* New York: Viking Penguin.

Ladefoged, P., and I. Maddieson. 1996. *The sounds of the world's languages.* Cambridge, Mass.: Blackwell.

Lane, H. L. 1976. *The wild boy of Aveyron.* Cambridge: Harvard University Press.

Lasky, R. E., A. Syrdal-Lasky, and R. E. Klein. 1975. VOT discrimination by four to six and a half month old infants from Spanish environments. *Journal of Experimental Child Psychology* 20:215–25.

Lempers, J. D., E. R. Flavell, and J. H. Flavell. 1977. The development in very young children of tacit knowledge concerning visual perception. *Genetic Psychology Monographs* 95:3–53.

Leslie, A. M. 1984. Spatiotemporal continuity and the perception of causality in infants. *Perception* 13:287–305.

———. 1987. Pretense and representation: The origins of "theory of mind." *Psychological Review* 94:412–26.

Leslie, A. M., and S. Keeble. 1987. Do six-month-old infants perceive causality? *Cognition* 25:265–88.

Levelt, W. J. M. 1989. *Speaking: From intention to articulation.* Cambridge: MIT Press.

Lewis, D. 1976. Observations on route finding and spatial orientation among aboriginal peoples of the western desert region of central Australia. *Oceania* 46:249–82.

Lewkowicz, D. J., and R. Lickliter, eds. 1994. *The development of intersensory perception: Comparative perspectives.* Hillsdale, N.J.: Erlbaum.

Liberman, A. M., F. S. Cooper, D. P. Shankweiler, and M. Studdert-Kennedy. 1967. Perception of the speech code. *Psychological Review* 74:431–61.

Lieberman, P. 1991. *Uniquely human: The evolution of speech, thought and selfless behavior.* Cambridge: Harvard University Press.

Lifter, K., and L. Bloom. 1989. Object knowledge and the emergence of language. *Infant Behavior and Development* 12:395–423.

Lillard, A. 1998. Ethnopsychologies: Cultural variations in theories of mind. *Psychological Bulletin* 123:3–32.

Lipsitt, L. P. 1969. Learning capacities of the human infant. In *Brain and early behaviour: Development in the fetus and infant,* ed. R. J. Robinson, 227–49. London: Academic Press.

———. 1978. A coming out occasion for babies. *Infant Behavior and Development* 1:1–2.

———. 1998. Interview by A. N. Meltzoff, April 7, included in the Society for Research in Child Development Oral History Project directed by W. W. Hartup. Tape recording, SRCD Executive Office, University of Michigan, Ann Arbor.

Lipsitt, L. P., and J. S. Werner. 1981. The infancy of human learning processes. In *Developmental plasticity,* ed. E. S. Gollin, 101–33. New York: Academic Press.

Lisker, L., and A. S. Abramson. 1964. A cross-language study of voicing in initial stops: Acoustical measurements. *Word* 20:384–422.

Locke, J. 1959. *An essay concerning human understanding.* New York: Dover Publications.

Locke, J. L. 1993. *The child's path to spoken language.* Cambridge: Harvard University Press.

Logan, J. S., S. E. Lively, and D. B. Pisoni. 1991. Training Japanese listeners to identify English /r/ and /l/: A first report. *Journal of the Acoustical Society of America* 89:874–86.

Lorenz, K. Z. 1937. The companion in the bird's world. *Auk* 54:245–73.

———. 1943. Die angeborenen Formen möglicher Erfahrung. *Zeitschrift für Tierpsychologie* 5:235–409.

Lycan, W. G. 1996. *Consciousness and experience.* Cambridge: MIT Press.

Macfarlane, A. 1975. Olfaction in the development of social preferences in the human neonate. *Ciba Foundation Symposium* 33:103–13.

MacKain, K., M. Studdert-Kennedy, S. Spieker, and D. Stern. 1983. Infant intermodal speech perception is a left-hemisphere function. *Science* 219:1347–49.

MacWhinney, B., and C. Snow. 1985. The child language data exchange system. *Journal of Child Language* 12:271–95.

———. 1990. The child language data exchange system: An update. *Journal of Child Language* 17:457–72.

Main, M. 1991. Metacognitive knowledge, metacognitive monitoring, and singular (coherent) vs. multiple (incoherent) model of attachment: Findings and directions for future research. In *Attachment across the life cycle,* ed. C. M. Parkes, J. Stevenson-Hinde, and P. Marris, 127–59. London: Tavistock/Routledge.

Markman, E. M. 1989. *Categorization and naming in children: Problems of induction.* Cambridge: MIT Press.

Marler, P. 1970a. A comparative approach to vocal learning: Song development in white-crowned sparrows. *Journal of Comparative and Physiological Psychology* 71:1–25.

———. 1970b. Birdsong and speech development: Could there be parallels? *American Scientist* 58:669–73.

Marr, D. 1982. *Vision: A computational investigation into the human representation and processing of visual information.* San Francisco: W. H. Freeman.

Masangkay, Z., K. McClusky, C. McIntyre, J. Sims-Knight, B. Vaughn, and J. H. Flavell. 1974. The early development inferences about the visual percepts of others. *Child Development* 45:357–66.

McQueen, J. M., D. Norris, and A. Cutler. 1994. Competition in spoken word recognition: Spotting words in other words. *Journal of Experimental Psychology: Learning, Memory, and Cognition* 20:621–38.

Medin, D. L., and L. W. Barsalou. 1987. Categorization processes and categorical perception. In *Categorical perception: The groundwork of cognition,* ed. S. Harnad, 455–90. New York: Cambridge University Press.

Mehler, J., and R. Fox, eds. 1985. *Neonate cognition: Beyond the blooming buzzing confusion.* Hillsdale, N.J.: Erlbaum.

Mehler, J., P. Jusczyk, G. Lambertz, N. Halsted, J. Bertoncini, and C. Amiel-Tison. 1988. A precursor of language acquisition in young infants. *Cognition* 29:143–78.

Meltzoff, A. N. 1988a. Infant imitation after a 1-week delay: Long-term memory for novel acts and multiple stimuli. *Developmental Psychology* 24:470–76.

———. 1988b. Infant imitation and memory: Nine-month-olds in immediate and deferred tests. *Child Development* 59:217–25.

———. 1988c. Imitation of televised models by infants. *Child Development* 59:1221–29.

———. 1988d. Imitation, objects, tools, and the rudiments of language in human ontogeny. *Human Evolution* 3:45–64.

———. 1990a. Foundations for developing a concept of self: The role of imitation in relating self to other and the value of social mirroring, social modeling, and self practice in infancy. In *The self in transition: Infancy to childhood,* ed. D. Cicchetti and M. Beeghly, 139–64. Chicago: University of Chicago Press.

———. 1990b. Towards a developmental cognitive science: The implications of cross-modal matching and imitation for the development of representation and memory in infancy. In *Annals of the New York Academy of Sciences: The development and neural bases of higher cognitive functions,* ed. A. Diamond, vol. 608, 1–31. New York: Annals of the New York Academy of Sciences.

———. 1995a. Understanding the intentions of others: Re-enactment of intended acts by 18-month-old children. *Developmental Psychology* 31:838–50.

———. 1995b. What infant memory tells us about infantile amnesia: Long-term recall and deferred imitation. *Journal of Experimental Child Psychology* 59:497–515.

Meltzoff, A. N., and R. W. Borton. 1979. Intermodal matching by human neonates. *Nature* 282:403–4.

Meltzoff, A. N., and A. Gopnik. 1993. The role of imitation in understanding persons and developing a theory of mind. In *Understanding other minds: Perspectives from autism,* ed. S. Baron-Cohen, H. Tager-

Flusberg, and D. J. Cohen, 335–66. New York: Oxford University Press.

Meltzoff, A. N., A. Gopnik, and B. M. Repacholi. 1999. Toddlers' understanding of intentions, desires, and emotions: Explorations of the dark ages. In *Development of intention and intentional understanding in infancy and early childhood,* ed. P. D. Zelazo, J. W. Astington, and D. R. Olson, 17–41. Mahwah, N.J.: Erlbaum.

Meltzoff, A. N., and M. K. Moore. 1977. Imitation of facial and manual gestures by human neonates. *Science* 198:75–78.

———. 1983. Newborn infants imitate adult facial gestures. *Child Development* 54:702–9.

———. 1992. Early imitation within a functional framework: The importance of person identity, movement, and development. *Infant Behavior and Development* 15:479–505.

———. 1994. Imitation, memory, and the representation of persons. *Infant Behavior and Development* 17:83–99.

———. 1995. Infants' understanding of people and things: From body imitation to folk psychology. In *Body and the self,* ed. J. Bermúdez, A. J. Marcel, and N. Eilan, 43–69. Cambridge: MIT Press.

———. 1997. Explaining facial imitation: A theoretical model. *Early Development and Parenting* 6:179–92.

———. 1998. Object representation, identity, and the paradox of early permanence: Steps toward a new framework. *Infant Behavior and Development* 21:201–35.

———. 1999a. Persons and representation: Why infant imitation is important for theories of human development. In *Imitation in infancy,* ed. J. Nadel and G. Butterworth, 9–35. Cambridge: Cambridge University Press.

———. 1999b. A new foundation for cognitive development in infancy: The birth of the representational infant. In *Conceptual development: Piaget's legacy,* ed. E. Scholnick, K. Nelson, P. Miller, and S. Gelman, 53–78. Mahwah, N.J.: Erlbaum.

Mervis, C. B. 1987. Child-basic object categories and early lexical development. In *Concepts and conceptual development: Ecological and intellectual factors in categorization,* ed. U. Neisser, 201–34. New York: Cambridge University Press.

Mervis, C. B., and J. Bertrand. 1994. Acquisition of the novel name-nameless category (N3C) principle. *Child Development* 65:1646–62.

———. 1997. Developmental relations between cognition and language: Evidence from Williams syndrome. In *Communication and lan-*

guage acquisition: Discoveries from atypical development, ed. L. B. Adamson and M. A. Romski. Baltimore: Brookes.

Mervis, C. B., and K. E. Johnson. 1991. Acquisition of the plural morpheme: A case study. *Developmental Psychology* 27:222–35.

Mervis, C. B., and J. R. Pani. 1980. Acquisition of basic object categories. *Cognitive Psychology* 12:496–522.

Mervis, C. B., and E. Rosch. 1981. Categorization of natural objects. *Annual Review of Psychology* 32:89–115.

Merzenich, M. M., W. M. Jenkins, P. Johnston, C. Schreiner, S. L. Miller, and P. Tallal. 1996. Temporal processing deficits of language-learning impaired children ameliorated by training. *Science* 271:77–81.

Middleton, B., M. A. Shwe, D. E. Heckerman, M. Henrion, E. Horitz, H. Lehmann, and G. Cooper. 1991. Probabilistic diagnosis using a reformulation of the INTERNIST-1/QMR knowledge base. Evaluation of diagnostic performance. *Methods of Information in Medicine* 30: 256–67.

Mielke, R., and W. D. Heiss. 1998. Positron emission tomography for diagnosis of Alzheimer's disease and vascular dementia. *Journal of Neural Transmission,* Supplementum 53:237–50.

Miller, G. A. 1996. *The science of words.* San Francisco: W. H. Freeman.

Miller, J. L. 1994. On the internal structure of phonetic categories: A progress report. *Cognition* 50:271–85.

Miller, R. A., H. E. Pople, and J. D. Myers. 1982. Internist-1: An experimental computer-based diagnostic consultant for general internal medicine. *New England Journal of Medicine* 307:468–76.

Miyawaki, K., W. Strange, R. Verbrugge, A. M. Liberman, J. J. Jenkins, and O. Fujimura. 1975. An effect of linguistic experience: The discrimination of [r] and [l] by native speakers of Japanese and English. *Perception and Psychophysics* 18:331–40.

Moon, C., R. P. Cooper, and W. P. Fifer. 1993. Two-day-olds prefer their native language. *Infant Behavior and Development* 16:495–500.

Moore, M. K., R. Borton, and B. L. Darby. 1978. Visual tracking in young infants: Evidence for object identity or object permanence? *Journal of Experimental Child Psychology* 25:183–98.

Moore, M. K., and A. N. Meltzoff. 1999. New findings on object permanence: A developmental difference between two types of occlusion. *British Journal of Developmental Psychology* 17.

Morel, A., P. E. Garraghty, and J. H. Kaas. 1993. Tonotopic organization, architectonic fields and connections of auditory cortex in macaque monkeys. *Journal of Comparative Neurology* 335:437–59.

Morgan, J. L., and K. Demuth, eds. 1995. *Signal to syntax: Bootstrapping from speech to grammar in early acquisition.* Hillsdale, N.J.: Erlbaum.

Morris, J. S., C. D. Frith, D. I. Perrett, and D. Rowland. 1996. A differential neural response in the human amygdala to fearful and happy facial expressions. *Nature* 383:812–15.

Morrongiello, B. A. 1994. Effects of colocation on auditory-visual interactions and cross-modal perception in infants. In *The development of intersensory perception: Comparative perspectives,* ed. D. J. Lewkowicz and R. Lickliter, 235–63. Hillsdale, N.J.: Erlbaum.

Morton, A. 1980. *Frames of mind: Constraints on the common-sense conception of the mental.* New York: Oxford University Press.

Muir, D. W., and S. M. J. Hains. 1993. Infant sensitivity to perturbations in adult facial, vocal, tactile, and contingent stimulation during face-to-face interactions. In *Developmental neurocognition: Speech and face processing in the first year of life,* ed. B. de Boysson-Bardies, S. de Schonen, P. Jusczyk, P. MacNeilage, and J. Morton, 171–85. Dordrecht, Netherlands: Kluwer.

Munakata, Y., J. L. McClelland, M. H. Johnson, and R. S. Siegler. 1997. Rethinking infant knowledge: Toward an adaptive process account of successes and failures in object permanence tasks. *Psychological Review* 104:686–713.

Murdock, G. P. 1959. Cross-language parallels in parental kin terms. *Anthropological Linguistics* 1:1–5.

Näätänen, R., A. Lehtokoski, M. Lennes, M. Cheour, M. Huotilainen, A. Iivonen, M. Vainio, P. Alku, R. J. Ilmoniemi, A. Luuk, J. Allik, J. Sinkkonen, and K. Alho. 1997. Language-specific phoneme representations revealed by electric and magnetic brain responses. *Nature* 385:432–34.

Nadel, J., and G. Butterworth. 1999. *Imitation in infancy.* Cambridge: Cambridge University Press.

Nadel-Brulfert, J., and P. M. Baudonnière. 1982. The social function of reciprocal imitation in 2-year-old peers. *International Journal of Behavioral Development* 5:95–109.

Naigles, L. G., and S. A. Gelman. 1995. Overextensions in comprehension and production revisited: Preferential-looking in a study of dog, cat, and cow. *Journal of Child Language* 22:19–46.

Nass, R. D., and S. Gazzaniga. 1985. Cerebral lateralization and specialization in human central nervous system. In *Handbook of Physiology,* ed. F. Plum, 701–61. Bethesda, Md.: The American Physiological Society.

Nelson, C. A. 1987. The recognition of facial expressions in the first two years of life: Mechanism of development. *Child Development* 58: 889–909.

Nelson, K. 1981. Individual differences in language development: Implications for development and language. *Developmental Psychology* 17: 170–87.

———. 1985. *Making sense: The acquisition of shared meaning.* New York: Academic Press.

———. 1990. Remembering, forgetting, and childhood amnesia. In *Knowing and remembering in young children. Emory symposia in cognition,* ed. R. Fivush and J. A. Hudson, vol. 3, 301–6. New York: Cambridge University Press.

———. 1996. *Language in cognitive development: The emergence of the mediated mind.* Cambridge: Cambridge University Press.

Neurath, O. 1959. Protocol sentences. In *Logical positivism,* ed. A. J. Ayer, 199–208. Glencoe, Ill.: Free Press.

Newcombe, N., and N. A. Fox. 1994. Infantile amnesia: Through a glass darkly. *Child Development* 65:31–40.

Newport, E. L. 1990. Maturational constraints on language learning. *Cognitive Science* 14:11–28.

———. 1991. Contrasting conceptions of the critical period for language. In *Epigenesis of mind: Essays on biology and cognition,* ed. S. Carey and R. Gelman, 111–30. Hillsdale, N.J.: Erlbaum.

Nottebohm, F. 1969. The "critical period" for song learning. *Behavioral and Neural Biology* 111:386–87.

Nottebohm, F., M. E. Nottebohm, and L. Crane. 1986. Developmental and seasonal changes in canary song and their relation to changes in the anatomy of song-control nuclei. *Behavioral and Neural Biology* 46:445–71.

Nygaard, L. C., and D. B. Pisoni. 1995. Speech perception: New directions in research and theory. In *Handbook of perception and cognition,* vol. 11, *Speech, language, and communication,* ed. J. L. Miller and P. D. Eimas, 63–96. San Diego: Academic Press.

O'Neill, D. K. 1996. Two-year-old children's sensitivity to a parent's knowledge state when making requests. *Child Development* 67:659–77.

O'Neill, D. K., J. W. Astington, and J. H. Flavell. 1992. Young children's understanding of the role that sensory experiences play in knowledge acquisition. *Child Development* 63:474–90.

O'Neill, D. K., and A. Gopnik. 1991. Young children's ability to identify the sources of their beliefs. *Developmental Psychology* 27:390–97.

Oakes, L. M., and L. B. Cohen. 1995. Infant causal perception. In *Advances in infancy research,* ed. C. Rovee-Collier and L. P. Lipsitt, vol. 9, 1–54. Norwood, N.J.: Ablex.

Ojemann, G. A. 1983. Brain organization for language from the perspective of electrical stimulation mapping. *Behavioral and Brain Sciences* 6:189–230.

Oller, D. K., and M. P. Lynch. 1992. Infant vocalizations and innovations in infraphonology: Toward a broader theory of development and disorders. In *Phonological development: Models, research, implications,* ed. C. A. Ferguson, L. Menn, and C. Stoel-Gammon, 509–36. Timonium, Md.: York.

Oyama, S. 1976. A sensitive period for the acquisition of a nonnative phonological system. *Journal of Psycholinguistic Research* 5:261–83.

Palmer, S. E. 1999. *Vision science: Photons to phenomenology.* Cambridge: MIT Press.

Papousek, H. 1969. Individual variability in learned responses in human infants. In *Brain and early behavior: Development in the fetus and infant,* ed. R. J. Robinson, 251–66. New York: Academic Press.

Papousek, H., and M. Papousek. 1984. Learning and cognition in the everyday life of human infants. In *Advances in the study of behavior,* ed. J. Rosenblatt, C. Beer, C. Busnel, and P. Slater, vol. 14, 127–63. New York: Academic Press.

Pascalis, O., S. de Schonen, J. Morton, C. Deruelle, and M. Fabre-Grenet. 1995. Mother's face recognition by neonates: A replication and an extension. *Infant Behavior and Development* 18:79–85.

Perner, J. 1991. *Understanding the representational mind.* Cambridge: MIT Press.

Perner, J., S. R. Leekam, and H. Wimmer. 1987. Three-year-olds' difficulty with false belief: The case for a conceptual deficit. *British Journal of Developmental Psychology* 5:125–37.

Perner, J., and T. Ruffman. 1995. Episodic memory and autonoetic consciousness: Developmental evidence and a theory of childhood amnesia. *Journal of Experimental Child Psychology* 59:516–48.

Perner, J., T. Ruffman, and S. R. Leekam. 1994. Theory of mind is contagious: You catch it from your sibs. *Child Development* 65:1228–38.

Perrett, D. I., M. H. Harries, A. J. Mistlin, J. K. Hietanen, P. J. Benson, R. Bevan, S. Thomas, M. W. Oram, J. Ortega, and K. Brierley. 1990. Social signals analyzed at the single cell level: Someone is looking at me, something touched me, something moved! *International Journal of Comparative Psychology* 4:25–55.

Perrett, D. I., J. K. Heitanen, M. W. Oram, and P. J. Benson. 1992. Organization and functions of cells responsive to faces in the temporal cortex. In *Processing the facial image,* ed. V. Bruce and A. Cowey, 23–30. Oxford: Clarendon.

Perrett, D. I., A. J. Mistlin, and A. J. Chitty. 1987. Visual neurons responsive to faces. *Trends in Neuroscience* 10:358–64.

Peskin, J. 1992. Ruse and representations: On children's ability to conceal information. *Developmental Psychology* 28:84–89.

Petersen, S. E., and J. A. Fiez. 1993. The processing of single words studied with positron emission tomography. *Annual Review of Neuroscience* 16:509–30.

Petersen, S. E., P. T. Fox, A. Z. Snyder, and M. E. Raichle. 1990. Activation of extrastriate and frontal cortical areas by visual words and word-like stimuli. *Science* 249:1041–44.

Petitto, L. A. 1993. On the ontogenetic requirements for early language acquisition. In *Developmental neurocognition: Speech and face processing in the first year of life,* ed. B. de Boysson-Bardies, S. de Schonen, P. Jusczyk, P. McNeilage, and J. Morton, 365–83. Dordrecht, Netherlands: Kluwer.

Piaget, J. 1952a. Jean Piaget. In *History of psychology in autobiography,* ed. E. G. Boring, H. S. Langfeld, H. Werner, and R. M. Yerkes, vol. 4, 237–40. Worcester, Mass.: Clark University Press.

———. 1952b. *The origins of intelligence in children.* New York: International Universities Press.

———. 1954. *The construction of reality in the child.* New York: Basic Books.

———. 1962. *Play, dreams and imitation in childhood.* New York: Norton.

Pinker, S. 1984. *Language learnability and language development.* Cambridge: Harvard University Press.

———. 1987. The bootstrapping problem in language acquisition. In *Mechanisms of language acquisition,* ed. B. MacWhinney, 399–441. Hillsdale, N.J.: Erlbaum.

———. 1994. *The language instinct.* New York: Morrow.

———. 1997. *How the mind works.* New York: Norton.

Plato. 1937a. *Meno.* In *The dialogues of Plato.* Trans. B. Jowett. New York: Random House.

———. 1937b. *Phaedo.* In *The dialogues of Plato.* Trans. B. Jowett. New York: Random House.

———. 1951. *The symposium.* Trans. W. Hamilton. Harmondsworth, England: Penguin.

Popper, K. R. 1965. *Conjectures and refutations: The growth of scientific knowledge.* 2nd ed. New York: Basic Books.

Porter, R. H., M. W. Makin, L. B. Davis, and K. M. Christensen. 1991. An assessment of the salient olfactory environment of formula-fed infants. *Physiology and Behavior* 50:907–11.

Posner, M. I., and S. Keele. 1970. Retention of abstract ideas. *Journal of Experimental Psychology* 83:304–8.

Posner, M. I., and M. E. Raichle. 1994. *Images of mind.* New York: W. H. Freeman.

Povinelli, D. J., and T. J. Eddy. 1996. What young chimpanzees know about seeing. *Monographs of the Society for Research in Child Development* 61, no. 3 (serial no. 247).

Povinelli, D. J., and T. M. Preuss. 1995. Theory of mind: Evolutionary history of a cognitive specialization. *Trends in Neurosciences* 18: 418–24.

Putnam, H. 1975. *Mind, language and reality: Philosophical papers.* Vol. 2. New York: Cambridge University Press.

Pylyshyn, Z. W. 1984. *Computation and cognition: Toward a foundation for cognitive science.* Cambridge: MIT Press.

Quine, W. V. O. 1960. *Word and object.* Cambridge: MIT Press.

Quinn, P. C., and P. D. Eimas. 1996. Perceptual organization and categorization in young infants. In *Advances in infancy research,* ed. C. Rovee-Collier and L. P. Lipsitt, vol. 10, 1–36. Norwood, N.J.: Ablex.

Repacholi, B. M. 1998. Infants' use of attentional cues to identify the referent of another person's emotional expression. *Developmental Psychology* 34:1017–25.

Repacholi, B. M., and A. Gopnik. 1997. Early reasoning about desires: Evidence from 14- and 18-month-olds. *Developmental Psychology* 33: 12–21.

Reznick, J. S., and B. A. Goldfield. 1992. Rapid change in lexical development in comprehension and production. *Developmental Psychology* 28:406–13.

Ricciuti, H. N. 1965. Object grouping and selective ordering behaviors in infants 12 to 24 months old. *Merrill-Palmer Quarterly* 11:129–48.

Rips, L. J. 1975. Inductive judgments about natural categories. *Journal of Verbal Learning and Verbal Behavior* 14:665–81.

Ritchie, D. 1986. *The computer pioneers: The making of the modern computer.* New York: Simon and Schuster.

Rizzolatti, G., and M. A. Arbib. 1998. Language within our grasp. *Trends in Neuroscience* 21:188–94.

Rizzolatti, G., L. Fadiga, V. Gallese, and L. Fogassi. 1996. Premotor cortex and the recognition of motor actions. *Brain Research* 3:131–41.

Rogoff, B. 1990. *Apprenticeship in thinking: Cognitive development in social context.* New York: Oxford University Press.

———. 1998. Cognition as a collaborative process. In *Handbook of child psychology,* ed. W. Damon. Vol. 2, *Cognition, perception, and language,* ed. D. Kuhn and R. Siegler, 679–744. New York: Wiley.

Rogoff, B., M. J. Sellers, S. Pirrotta, N. Fox, and S. White. 1975. Age of assignment of roles and responsibilities to children: A cross-cultural survey. *Human Development* 18:353–69.

Rosch, E. 1975. Cognitive reference points. *Cognitive Psychology* 7:532–47.

Rousseau, J. J. 1974. *Émile.* Trans. B. Foxley. London: Dent.

Rovee-Collier, C. K. 1990. The "memory system" of prelinguistic infants. In *Annals of the New York Academy of Sciences: The development and neural bases of higher cognitive functions,* ed. A. Diamond, vol. 608, 517–42. New York: New York Academy of Sciences.

Rovee-Collier, C. K., and M. J. Gekoski. 1979. The economics of infancy: A review of conjugate reinforcement. In *Advances in child development and behavior,* ed. H. W. Reese and L. P. Lipsitt, vol. 14, 195–255. New York: Academic Press.

Rovee-Collier, C. K., and L. P. Lipsitt. 1982. Learning, adaptation, and memory in the newborn. In *Psychobiology of the human newborn,* ed. P. Stratton, 147–90. New York: Wiley.

Rovee-Collier, C. K., M. W. Sullivan, M. Enright, D. Lucas, and J. W. Fagen. 1980. Reactivation of infant memory. *Science* 208:1159–61.

Ruffman, T., J. Perner, M. Naito, L. Parkin, and W. A. Clements. 1998. Older (but not younger) siblings facilitate false belief understanding. *Developmental Psychology* 34:161–74.

Russell, B. 1905. On denoting. *Mind* 14:479–93.

———. 1948. Science as a product of Western Europe. *The Listener* 39: 865–66.

Russell, J., C. Jarrold, and D. Potel. 1994. What makes strategic deception difficult for children—the deception or the strategy? *British Journal of Developmental Psychology* 12:301–14.

Ryle, G. 1949. *The concept of mind.* New York: Harper & Row.

Sacks, O. 1995. *An anthropologist on Mars.* New York: Vintage.

Saffran, J. R., R. N. Aslin, and E. L. Newport. 1996. Statistical learning by 8-month-old infants. *Science* 274:1926–28.

Salapatek, P., and L. Cohen, eds. 1987. *Handbook of infant perception: From perception to cognition.* New York: Academic Press.

Scarr, S. 1998. American child care today. *American Psychologist* 53:95–108.

Searle, J. R. 1984. *Minds, brains and science.* Cambridge: Harvard University Press.

Shankle, W. R., B. H. Landing, M. S. Rafii, A. Schiano, J. M. Chen, and J. Hara. 1998. Evidence for a postnatal doubling of neuron number in the developing human cerebral cortex between 15 months and 6 years. *Journal of Theoretical Biology* 191:115–40.

Shatz, C. J. 1990. Impulse activity and the patterning of connections during CNS development. *Neuron* 5:745–56.

———. 1992. The developing brain. *Scientific American* 267:61–67.

Shatz, M., and R. Gelman. 1973. The development of communication skills: Modifications in the speech of young children as a function of listener. *Monographs of the Society for Research in Child Development* 38, no. 5 (serial no. 152).

Shore, B. 1996. *Culture in mind: Cognition, culture, and the problem of meaning.* New York: Oxford University Press.

Shumeiko, N. S. 1998. Age-related changes in the cytoarchitectonics of the human sensorimotor cortex. *Neuroscience and Behavioral Physiology* 28:345–48.

Shweder, R. 1991. *Thinking through cultures: Expeditions in cultural psychology.* Cambridge: Harvard University Press.

Shweder, R., J. Goodnow, G. Hatano, R. LeVine, H. Markus, and P. Miller. 1998. The cultural psychology of development: One mind, many mentalities. In *Handbook of child psychology,* ed. W. Damon. Vol. 1, *Theoretical models of human development,* ed. R. M. Lerner, 865–937. New York: Wiley.

Siegler, R. S. 1998. *Children's thinking.* 3d ed. Upper Saddle River, N.J.: Prentice-Hall.

Sigman, M., and L. Capps, eds. 1997. *Children with autism: A developmental perspective. The developing child.* Cambridge: Harvard University Press.

Simonds, R. J., and A. B. Scheibel. 1989. The postnatal development of the motor speech area: A preliminary study. *Brain and Language* 37:42–58.

Skinner, B. F. 1948. *Walden two.* New York: Macmillan.

———. 1971. *Beyond freedom and dignity.* New York: Alfred A. Knopf.

Skolnick, A. S., and J. H. Skolnick. 1992. *Family in transition: Rethinking marriage, sexuality, child rearing, and family organization.* 7th ed. New York: HarperCollins.

Slater, A., V. Morison, and D. Rose. 1984. Habituation in the newborn. *Infant Behavior and Development* 7:183–200.

Slater, A., A. Mattock, and E. Brown. 1990. Size constancy at birth: Newborn infants' responses to retinal and real size. *Journal of Experimental Child Psychology* 49:314–22.

Slaughter, V., and A. Gopnik. 1996. Conceptual coherence in the child's theory of mind: Training children to understand belief. *Child Development* 67:2967–88.

Slobin, D. I., ed. 1992–1997. *The crosslinguistic study of language acquisition.* 5 vols. Hillsdale, N.J.: Erlbaum.

Snow, C. E. 1977. The development of conversation between mothers and babies. *Journal of Child Language* 4:1–22.

———. 1987. Relevance of the notion of a critical period to language acquisition. In *Sensitive periods in development: Interdisciplinary perspectives,* ed. M. H. Bornstein, 183–209. Hillsdale, N.J.: Erlbaum.

Snow, C. E., and C. A. Ferguson. 1977. *Talking to children: Language input and acquisition.* New York: Cambridge University Press.

Sodian, B. 1991. The development of deception in young children. *British Journal of Developmental Psychology.* Special issue, *Perspectives on the child's theory of mind: I* 9:173–88.

Sodian, B., C. Taylor, P. L. Harris, and J. Perner. 1991. Early deception and the child's theory of mind: False trails and genuine markers. *Child Development* 62:468–83.

Spelke, E. S. 1979. Perceiving bimodally specified events in infancy. *Developmental Psychology* 15:626–36.

———. 1987. The development of intermodal perception. In *Handbook of infant perception,* ed. P. Salapatek and L. Cohen. Vol. 2, *From perception to cognition,* 233–73. New York: Academic Press.

———. 1998. Nativism, empiricism, and the origins of knowledge. *Infant Behavior and Development* 21:181–200.

Spelke, E. S., K. Breinlinger, K. Jacobson, and A. Phillips. 1993. Gestalt relations and object perception: A developmental study. *Perception* 22: 1483–1501.

Spelke, E. S., K. Breinlinger, J. Macomber, and K. Jacobson. 1992. Origins of knowledge. *Psychological Review* 99:605–32.

Spelke, E. S., and E. L. Newport. 1998. Nativism, empiricism, and the development of knowledge. In *Handbook of child psychology,* ed. W. Damon. Vol. 1, *Theoretical models of human development,* ed. R. M. Lerner, 275–340. New York: Wiley.

Sperber, D. 1996. *Explaining culture: A naturalistic approach.* Oxford: Blackwell.

Springer, K. 1996. Young children's understanding of a biological basis for parent–offspring relations. *Child Development* 67:2841–56.

Springer, K., and F. C. Keil. 1989. On the development of biologically specific beliefs: The case of inheritance. *Child Development* 60: 637–48.

———. 1991. Early differentiation of causal mechanisms appropriate to biological and nonbiological kinds. *Child Development* 62:767–81.

Stager, C., and J. Werker. 1997. Infants listen for more phonetic detail in speech perception than in word-learning tasks. *Natural* 388:381–382.

Stern, D. N. 1985. *The interpersonal world of the infant: A view from psychoanalysis and developmental psychology.* New York: Basic Books.

Stern, D. N., S. Spieker, R. K. Barnett, and K. MacKain. 1983. The prosody of maternal speech: Infant age and context related changes. *Journal of Child Language* 10:1–15.

Stevens, K. N. 1998. *Acoustic phonetics.* Cambridge: MIT Press.

Stich, S. P. 1983. *From folk psychology to cognitive science: The case against belief.* Cambridge: MIT Press.

Strange, W., and S. Dittmann. 1984. Effects of discrimination training on the perception of /r-l/ by Japanese adults learning English. *Perception and Psychophysics* 36:131–45.

Strauss, M. S. 1979. Abstraction of prototypical information by adults and 10-month-old infants. *Journal of Experimental Psychology: Human Learning and Memory* 5:618–32.

Streeter, L. A. 1976. Language perception of 2-month-old infants shows effects of both innate mechanisms and experience. *Nature* 259:39–41.

Studdert-Kennedy, M., A. M. Liberman, K. S. Harris, and F. S. Cooper. 1970. Motor theory of speech perception: A reply to Lane's critical review. *Psychological Review* 77:234–49.

Studdert-Kennedy, M., and M. Mody. 1995. Auditory temporal perception deficits in the reading-impaired: A critical review of the evidence. *Psychonomic Bulletin and Review* 2:508–14.

Sugarman, S. 1983. *Children's early thought: Developments in classification.* New York: Cambridge University Press.

Sulloway, F. J. 1996. *Born to rebel: Birth order, family dynamics, and creative lives.* New York: Pantheon Books.

Tallal, P., M. M. Merzenich, S. Miller, and W. Jenkins. 1998. Language learning impairments: Integrating basic science, technology, and remediation. *Experimental Brain Research* 123:210–19.

Tallal, P., S. Miller, and R. H. Fitch. 1993. Neurobiological basis of speech: A case for the preeminence of temporal processing. In *Temporal information processing in the nervous system: Special reference to dyslexia and dysphasia,* ed. P. Tallal, A. M. Galaburda, R. R. Llinas, and C. V. Euler, 27–47. New York: The New York Academy of Sciences.

Tallal, P., S. L. Miller, G. Bedi, G. Byma, X. Wang, S. S. Nagarajan, C. Schreiner, W. M. Jenkins, and M. M. Merzenich. 1996. Language comprehension in language–learning impaired children improved with acoustically modified speech. *Science* 271:81–84.

Tardif, T., M. Shatz, and L. Naigles. 1997. Caregiver speech and children's use of nouns versus verbs: A comparison of English, Italian, and Mandarin. *Journal of Child Language* 24:535–65.

Taylor, M. 1996. A theory of mind perspective on social cognitive development. In *Handbook of perception and cognition,* ed. E. C. Carterette and M. P. Friedman. Vol. 13, *Perceptual and cognitive development,* ed. R. Gelman and T. Au, 283–329. New York: Academic Press.

Taylor, M., B. M. Esbensen, and R. T. Bennet. 1994. Children's understanding of knowledge acquisition: The tendency for children to report they have always known what they have just learned. *Child Development* 65:1581–1604.

Tomasello, M., and M. E. Barton. 1994. Learning words in nonostensive contexts. *Developmental Psychology* 30:639–50.

Tomasello, M., and J. Call. 1997. *Primate cognition.* New York: Oxford University Press.

Tomasello, M., and M. J. Farrar. 1986. Object permanence and relational words: A lexical training study. *Journal of Child Language* 13: 495–505.

Tomasello, M., A. C. Kruger, and H. H. Ratner. 1993. Cultural learning. *Behavioral and Brain Sciences* 16:495–552.

Tomasello, M., and W. E. Merriman, eds. 1995. *Beyond names for things: Young children's acquisition of verbs.* Hillsdale, N.J.: Erlbaum.

Tomasello, M., R. Strosberg, and N. Akhtar. 1996. Eighteen-month-old children learn words in non-ostensive contexts. *Journal of Child Language* 23:157–76.

Trevarthen, C. 1979. Communication and cooperation in early infancy: A description of primary intersubjectivity. In *Before speech,* ed. M. Bullowa, 321–47. New York: Cambridge University Press.

Trevarthen, C., and P. Hubley. 1978. Secondary intersubjectivity: Confidence, confiding and acts of meaning in the first year. In *Action, gesture, and symbol: The emergence of language,* ed. A. Lock, 183–229. New York: Academic Press.

Turing, A. M. 1950. Computing machinery and intelligence. *Mind* 59: 433–60.

Uzgiris, I. C., and J. M. Hunt. 1975. *Assessment in infancy: Ordinal scales of psychological development.* Urbana: University of Illinois Press.

Vargha-Khadem, F., L. J. Carr, E. Isaacs, E. Brett, C. Adams, and M. Mishkin. 1997. Onset of speech after left hemispherectomy in a nine-year-old boy. *Brain* 120:159–82.

Vihman, M. M., and B. de Boysson-Bardies. 1994. The nature and origins of ambient language influence on infant vocal production and early words. *Phonetica* 51:159–69.

Vygotsky, L. S. 1967. Play and its role in the mental development of the child. *Soviet Psychology* 5:6–18.

———. 1986. *Thought and language.* Trans. A. Kozulin. Cambridge: MIT Press.

Walker-Andrews, A. S. 1997. Infants' perception of expressive behaviors: Differentiation of multimodal information. *Psychological Bulletin* 121:437–56.

Walton, G. E., N. J. A. Bower, and T. G. R. Bower. 1992. Recognition of familiar faces by newborns. *Infant Behavior and Development* 15: 265–69.

Walton, G. E., and T. G. R. Bower. 1993. Amodal representations of speech in infants. *Infant Behavior and Development* 16:233–43.

Waters, E., B. E. Vaughn, G. Posada, and K. Kondo-Ikemura. 1995. Caregiving, cultural, and cognitive perspectives on secure-base behavior and working models: New growing points of attachment theory and research. *Monographs of the Society for Research in Child Development* 60: nos. 2–3 (serial no. 244).

Watson, J. B. 1928. *Psychological care of infant and child.* New York: Norton.

———. 1930. *Behaviorism.* Chicago: University of Chicago Press.

Watson, J. S. 1972. Smiling, cooing and the "game." *Merrill-Palmer Quarterly* 18:323–39.

Weber, R., and J. Crocker. 1983. Cognitive processes in the revision of stereotypic beliefs. *Journal of Personality and Social Psychology* 45: 961–77.

Wellington, N., and M. J. Rieder. 1993. Attitudes and practices regarding analgesia for newborn circumcision. *Pediatrics* 92:541–43.

Wellman, H. M. 1990. *The child's theory of mind.* Cambridge: MIT Press.

Wellman, H. M., and S. A. Gelman. 1992. Cognitive development: Foundational theories of core domains. *Annual Review of Psychology* 43:337–75.

———. 1998. Knowledge acquisition in foundational domains. In *Handbook of child psychology,* ed. W. Damon. Vol. 2, *Cognition, perception, and language,* ed. D. Kuhn and R. Siegler, 523–73. New York: Wiley.

Wellman, H. M., A. K. Hickling, and C. A. Schult. 1997. Young children's psychological, physical, and biological explanations. In *The emergence of core domains of thought: Children's reasoning about physical, psychological, and biological phenomena. New directions for child development,* no. 75, ed. H. M. Wellman and K. Inagaki, 7–25. San Francisco: Jossey-Bass.

Werker, J. 1991. The ontogeny of speech perception. In *Modularity and the motor theory of speech perception,* ed. I. G. Mattingly and M. Studdert-Kennedy, 91–109. Hillsdale, N.J.: Erlbaum.

Werker, J. F., and R. C. Tees. 1984. Cross-language speech perception: Evidence for perceptual reorganization during the first year of life. *Infant Behavior and Development* 7:49–63.

Werner, E. E., and R. S. Smith. 1998. *Vulnerable, but invincible: A longitudinal study of resilient children and youth.* New York: Adams, Bannister, Cox.

Wertheimer, M. 1961. Psychomotor coordination of auditory and visual space at birth. *Science* 134:1692.

Wertsch, J. V. 1985. *Vygotsky and the social formation of mind.* Cambridge: Harvard University Press.

Whiten, A. 1991. *Natural theories of mind: Evolution, development, and simulation of everyday mindreading.* Oxford: Basil Blackwell.

Willatts, P. 1984. The stage-IV infant's solution of problems requiring the use of supports. *Infant Behavior and Development* 7:125–34.

———. 1989. Development of problem-solving in infancy. In *Infant development,* ed. A. Slater and G. Bremner, 143–82. Hillsdale, N.J.: Erlbaum.

Wimmer, H., J. Hogrefe, and J. Perner. 1988. Children's understanding of informational access as source of knowledge. *Child Development* 59: 386–96.

Wimmer, H., and J. Perner. 1983. Beliefs about beliefs: Representation and constraining function of wrong beliefs in young children's understanding of deception. *Cognition* 13:103–28.

Winnicott, D. W. 1971. *Playing and reality.* London: Tavistock Publications.

Wittgenstein, L. 1953. *Philosophical investigations.* Trans. G. E. M. Anscombe. Oxford: Basil Blackwell.

Woodward, A. 1998. Infants selectively encode the goal of an actor's reach. *Cognition* 69:1–34.

Woodward, A. L., E. M. Markman, and C. M. Fitzsimmons. 1994. Rapid word learning in 13- and 18-month-olds. *Developmental Psychology* 30: 553–66.

Wordsworth, W. 1943. "Ode: Intimations of immortality from recollections of early childhood." In *Wordsworth: Selected poems,* ed. W. E. Williams. New York: Penguin.

Xu, F., and S. Carey. 1996. Infants' metaphysics: The case of numerical identity. *Cognitive Psychology* 30:111–53.

Yonas, A., and C. Owsley. 1987. Development of visual space perception. In *Handbook of infant perception,* ed. P. Salapatek and L. Cohen. Vol. 2, *From perception to cognition,* 79–122. Orlando: Academic.

Zahn-Waxler, C., M. Radke-Yarrow, E. Wagner, and M. Chapman. 1992. Development of concern for others. *Developmental Psychology* 28: 126–36.

Zatorre, R. J., A. C. Evans, E. Meyer, and A. Gjedde. 1992. Lateralization of phonetic and pitch discrimination in speech processing. *Science* 256:846–49.

INDEX

C

F

N

O

P

Q

R